NEURAL DATA SCIENCE

science &
technology books

ELSEVIER

Companion Web Site:

https://www.elsevier.com/books-and-journals/book-companion/9780128040430

Neural Data Science
Erik Lee Nylen and Pascal Wallisch

TOOLS FOR ALL YOUR TEACHING NEEDS
textbooks.elsevier.com

ACADEMIC PRESS

NEURAL DATA SCIENCE

A Primer with MATLAB® and Python™

Erik Lee Nylen
Parsec Media, New York, NY, United States

Pascal Wallisch
New York University, New York, NY, United States

ACADEMIC PRESS

An imprint of Elsevier
elsevier.com

British Library Cataloguing-in-Publication Data
A catalogue record for this book is available from the British Library

Library of Congress Cataloging-in-Publication Data
A catalog record for this book is available from the Library of Congress

ISBN: 978-0-12-804043-0

For Information on all Academic Press publications
visit our website at https://www.elsevier.com/books-and-journals

Working together
to grow libraries in
developing countries

www.elsevier.com • www.bookaid.org

Publisher: Mara Conner
Acquisition Editor: Natalie Farra
Editorial Project Manager: Kathy Padilla
Production Project Manager: Edward Taylor
Designer: Mark Rogers

Typeset by MPS Limited, Chennai, India

Printed in the United States of America

Last digit is the print number: 9 8 7 6 5 4

Dedication

Dedicated to our grandfathers,
Karl Rädle and Lee Nylen

Contents

I

FOUNDATIONS

1. Philosophy

2. From 0 to 0.01

II
NEURAL DATA ANALYSIS

3. Wrangling Spike Trains

4. Correlating Spike Trains

5. Analog Signals

6. Biophysical Modeling

III
GOING BEYOND THE DATA

7. Regression

8. Dimensionality Reduction

9. Classification and Clustering

10. Web Scraping

Biography

Erik Lee Nylen received his PhD from the Center for Neural Science at New York University, and his BSE and MS in Biomedical Engineering at the University of Iowa. He did a fellowship at Insight Data Science, and has taught at the Neural Data Science summer course at Cold Spring Harbor Laboratory. He is a patented inventor and has performed with numerous musical groups. He is currently a data scientist in New York, where he also is Executive Codirector of The Stand, the New York City Dance Marathon.

Pascal Wallisch serves as a professor in the Department of Psychology at New York University where he currently teaches statistics, programming, and the use of mathematical tools in neuroscience and psychology. He received his PhD in Psychology from the University of Chicago and worked as a postdoctoral fellow at the Center for Neural Science at New York University. He has a long-term commitment and is dedicated to educational excellence, which was recognized by the "Wayne C. Booth Graduate Student Prize for Excellence in teaching" at the University of Chicago and the "Golden Dozen Award" at New York University. He cofounded and coorganizes the "Neural Data Science" summer course at Cold Spring Harbor Laboratory and coauthored "Matlab for Neuroscientists."

Preface

The future of neuroscience is likely to be very different from its past. Even though the field of neuroscience is inherently interdisciplinary, it has been dominated by neurobiology, a subfield of biology. As a corollary, the field has been in a state of relative data scarcity up to this point. However, biology itself is becoming ever more quantitative, and so is neuroscience. This development is paralleled by an ever-increasing torrent of brain-derived data, due to the availability of large-scale data recording techniques. Put simply, neuroscientists of the future will routinely be expected to handle massive amounts of data. Given the current state of the field, we expect this transition to be challenging, as the ability to do so is not within the standard repertoire of neuroscience training. But it also offers tremendous opportunities—it has been pointed out that available neuroscience methods are in principle insufficient to understand the basic workings of even simple information-processing systems where ground truth is known (Jonas & Kording, 2016). The brain is vastly more complicated, perhaps not even primarily an information-processing device and came about as a result of complex evolutionary processes with unknown design principles. To have any hope of understanding this extremely complex organ (and the mind it gives rise to) neuroscience will have to change. Fortunately, our ability to record more and more data from the brain and the physiological processes going on within it is ever-increasing, as are our methods to analyze these data, albeit with some lag (Stevenson & Kording, 2011). Over time, a continuously rising share of the latent information inherent in ongoing neural processes will become available in the form of data to inform our creation of models and theories of how the brain works. It is critical that this development is met by a large cadre of neuroscientists who are ready and able to handle these big data.

That is where this book comes in. "Data Science," an emerging field that is dedicated to the understanding of patterns in large datasets can, and we believe should, be brought to bear on neural data. Therefore, we aim to introduce "Neural Data Science"—principles from data science applied to neural data (with all its inherent complexities and idiosyncrasies) to a broader audience of neuroscientists.

Thus, this book is aimed at people who want to get a start in this field, as we believe they will need to in order for continued professional success within the neuroscience of the future. In order to reach as broad an audience as possible, we deliberately do not presuppose anything, no prior experience with scientific programming, linear algebra, calculus, machine learning, or statistics. This approach (conceptualized as "0 to 1 teaching") is baked in throughout the book. All that the prospective reader (you, as you are reading this) is expected to possess is a burning desire to learn how to handle "big data" in neuroscience. On the flip-side, this book is explicitly not aimed at people who are already well versed in these topics. The stated purpose of this book is to build solid foundations—to introduce the most relevant concepts in as clear a fashion as possible—so that the reader can build whatever structure they wish on these foundations. We cannot anticipate specific idiosyncratic use cases nor where the field itself will be going in the long term, but the structure of knowledge works to our advantage: In general, knowledge tends to be organized in a fractal or tree-like structure (see cartoon below). A tree has roots, a trunk, branches, and leaves. Experts working on the forefront of knowledge are mostly concerned about leaves and might even lose an appreciation for foundational knowledge. In contrast, this is a book about roots and trunks—introducing this foundational knowledge and how it came about, although we'll also explore some branches and leaves that are of particular interest or exemplary. We believe that by the end of the book, you will be in a perfect position to go from 1 to 100, or beyond, whatever that might be for your particular use case, with confidence.

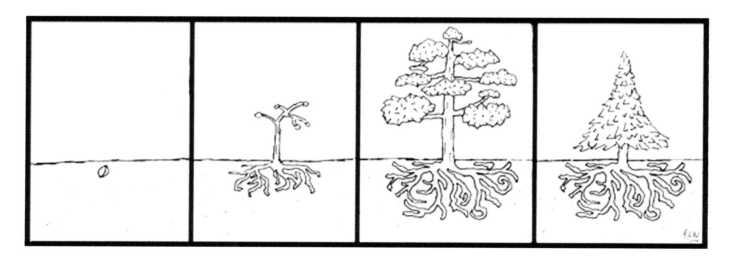

Specifically, this book is divided into four parts: In the first part, we'll explore philosophical underpinnings of the field and conceptual foundations that are truly elementary. In the second part, we cover techniques from a field that could properly be described as "computational neuroscience" from a modern perspective. This is important because this classical approach still dominates neuroscience today and you would be well advised to be familiar with it before moving on to the next level. The next level (Part 3) consists of the burgeoning data science techniques applied to neural data. The final part of the book comprises several appendices that are self-contained repositories of useful information that didn't fit any of the other parts.

One design decision we made concerns the fact that we are covering all material (wherever possible) in a practical fashion, with (commented) executable code in the two most relevant languages of scientific computing—MATLAB (still the most prevalent language in neuroscience) and Python (the most prevalent language in data science and becoming ever more relevant in neuroscience). In addition to this code, each chapter contains an exposition of what we are trying to accomplish—the strategic goal, why it is important, a discussion of several tactical approaches (algorithms) to achieve this goal, and practical considerations that arise as we implement these algorithms in code.

Writing a helpful book (as we hope this one will be) is a nontrivial, yet delightful, challenge and fundamentally a team effort. Thus, we are indebted to a bevy of individuals without whom we could not have hoped to succeed. First and foremost, we would like to emphasize the unwavering support we received from our respective families and friends. Second, we are indebted to Kathy Padilla, Natalie Farra, Mica Haley, and Edward Taylor, Calum Ross, Karen Giacomucci, Mark Rogers and Rob Duckwall at our publisher for their understanding, support, and patience. We would like to thank Julie Cachia for her helpful comments on the manuscript. We are grateful to Matthew A. Smith for sharing neural data as well as to Mike X Cohen and Mark Reimers for helpful feedback on an earlier version of this manuscript. We also would like to thank everyone else who has helped us along the way but whom we did not mention explicitly. Finally, we thank you—the reader—for the trust you place in us for considering to work through this book. We hope that we provide sufficient value to you in return. We would also like to stress that all remaining errors and shortcomings of this book are ours, and ours alone.

Neuroscience is undergoing radical, transformational change. We believe that the only way for neuroscience practitioners to prosper long term is if they embrace these developments, as they open up new and exciting ways to understand the brain. We hope that this book will serve as a useful guide for all those who are in the process of making this journey.

Pascal Wallisch
Erik Lee Nylen
New York, June 2016

How to Use This Book

As the content of this book is somewhat technical in nature, we want to be explicit about how to use it, so that you will be able to get the most out of it.

What is this book trying to do and who is it aimed at? The point of this book is to make the reader *code-safe*. This means that if you work through this book (either in the context of a course or by yourself), we believe that you will be able to explore any topic within the scope of neural data science you desire in more detail, without drowning. Actual programming proficiency will take thousands of hours of deliberate engagement with some problem area, actively coding. There are no good shortcuts to this, but the point of this book is to put in a couple of hundred carefully structured (and guided) hours so that you can explore with confidence, later.

The Rosetta Stone Approach. We use as inspiration the original Rosetta stone (Champollion, 1828). To make the book accessible to the largest possible audience, we decided to cover all materials in the most relevant languages in parallel, wherever possible. We recognize that the native language of most current neuroscientists will be Matlab, and we want to enable them to readily switch back and forth between Python and Matlab, should they desire to do so. Our language of choice for *pseudocode*—the words that describe what the code is supposed to do—is English. We also use English for the prose that specifies the strategic goals and the algorithms to implement them.

The pseudocode portion will have a dedicated panel, next to a panel that contains a figure resulting from the corresponding code (either Matlab or Python), where appropriate. This approach lets the reader learn Python from Matlab, Matlab from Python, Python from English, Matlab from English, or even learn English from prior knowledge of either Python or Matlab.

Python	Matlab	Pseudocode	Figure
>>> 5+ 7 12	>> 5+ 7 ans=12	Here we simply add five and seven and the output twelve is printed	No relevant figure here

Sometimes, where appropriate due to the spatial arrangements of the data in question, we will use a horizontal arrangement, like in the original Rosetta stone that dates from 196 BC:

Pseudocode	Sum up the numbers in the vector
Python	In >>> sum([0,0,0,0,0,0,0,0,0,1,0,1,0,1,0,0,0,1,0,0,0]) Out >>> 4
MATLAB	>> sum([0,0,0,0,0,0,0,0,0,1,0,1,0,1,0,0,0,1,0,0,0]) ans = 4

Fonts. The main text is set in Roman Palatino LT Std. Code input and output is set in `bold Courier New`. Figure captions are in arial 11

Function names. Any time we discuss the names of functions in the prose text, they will be italicized, like this: "To take the absolute value of a number, you can use the *abs* function."

Equations. Equations are all numbered by chapter number dot equation number and referred to as such in the prose text, e.g., "as you can see in equation 6.1, the output Y is a function of multiple parameters."

Pensees and cartoons. We will occasionally use *Pensees* to convey messages that do not necessarily fit into the main body of the text, but are additional reading from which Pascal offers his thoughts. These are deliberately designed not to mince words. Academics pride themselves on nuance. Everything is always hedged. The only good thing is that they only (at most) have two hands. They usually fail to understand that unless they are an octopus, they can't make an argument nuanced enough anyway. Sad. The part of the book that triggered the thought will be denoted by a ⌘, the Icelandic sign for "place of interest." We are appropriating it here to denote "point of interest." The cartoons (provided by Erik) are strategically used to illustrate concepts or ideas that benefit from a visual depiction.

Definitions. We provide a glossary of technical terms used across the many fields from which Neural Data Science draws in our *Glossary*. The Glossary will occasionally even include relevant terms that do not appear in the text. These are words that were deemed as useful to the reader wishing to understand basic definitions commonly used in the field even if it goes beyond the scope of the book. Generally speaking, if you don't know a term, look in the glossary for an explanation. Hopefully it will be there.

References. Throughout the book, we will reference a kaleidoscope of scholarly works from many academic fields. These references are organized (by chapter)—in the bibliography at the end of the book.

Companion website. All figures and executable code within this book can be downloaded from the companion website for this book. The information on how to access this repository is on page ii. If there should be a code error somewhere in the printed version of the book, the online code will—hopefully—provide a version where we fixed the error by the time you are reading this.

When the book is not enough. We encourage you to use resources such as Google, Stack Exchange, or Stack Overflow. Copying and pasting errors into Google is typically the best way to find solutions to your errors. Do not hesitate to do so, particularly as this field is rapidly evolving and any book on this topic risks obsolescence upon release.

PART I

FOUNDATIONS

1

Philosophy

Before we get started—and perhaps even bogged down—with specific algorithms and how to implement them in code, we want to lay out how we conceptualize the emerging field of neural data science (NDS) as well as the relevant challenges within neuroscience that NDS can help address. This is important so as to not get lost in the random forest.

WHAT IS DATA SCIENCE?

The field of data science emerged in the early 21st century as a distinct scientific discipline. It can be thought of as a fusion of concepts from computer science, statistics, and machine learning in conjunction with the availability of both an abundance of data as well as sufficient computational power that allows for the implementation of these concepts on this data, in code. (Hilbert & López, 2011). Data science has plenty of historical antecedents that can be traced well into the 20th century. However, people that were doing "data science" then would be surprised to learn that they were doing so at the time—instead, they saw themselves as doing statistics (Tukey, 1977), working on problems of communication theory (Shannon, 1948), computational theories (Turing, 1938), or cryptanalysis (Good, 1979), to name just a few.

The term "data scientist" (a practitioner of data science) was coined by Jeff Hammerbacher and D.J. Patil in 2008 (Watson, 2013). It has been pointed out that the term *data science* is somewhat unfortunate, as there is no science without data, in this sense every science is a data science (Wallisch, 2013, 2014). As will hopefully become clear shortly, "neural metralogy" would be a more suitable term. However, data science in the more narrow sense describes the relentless pursuit of understanding and explaining the structure of data, of making predictions from this understanding using any means necessary, mostly by using algorithms and principles from computer science, statistics, and machine learning. Data science explicitly transcends classical discipline boundaries, e.g., a biologist is usually only interested in data from biology,

Neural Data Science.
DOI: http://dx.doi.org/10.1016/B978-0-12-804043-0.00001-5

a chemist only interested in data relevant to questions in chemistry, and so on. In data science, the data themselves take center stage, within an explicitly data-centric framework. In its strong form, the source of the data does not matter, as data are data. The first time that a data scientist has been featured in popular fiction was, to our knowledge, in season 4 of the Netflix television show "House of Cards," though individuals such as Nate Silver have prominently—if not always successfully—provided insights into data from tribalist entertainment media, in particular, sports and politics. As of 2016, data science is still rapidly growing and changing, but a consensus about its core concepts has now been established.

WHAT IS NEURAL DATA SCIENCE?

If the statements in our previous section are true (we believe they are), *neural data science* appears to be a contradiction in terms. How can one have a general data science that is deliberately indifferent about the source of its data but is at the same time bounded by neural data? This paradox can be resolved: not all data are created equal. In our view, the source of the data—which determines its structure and quality—does matter when it comes to the most suitable tools for its analysis. Similarly, the questions asked in a particular discipline frame the data recording in the first place, so we should not pretend that the data fell from the sky (or, rather, the cloud). In neuroscience, data are hard won by dedicated experimenters with specific questions; we should not pretend otherwise. Put differently, the principles of data science can be applied to neural data, but the structure of neural data is so rich that it frames the application and in fact might push the development of data science itself. Historically, mathematical frameworks were created to understand specific physical phenomena for instance, *calculus* was developed to understand planetary motion (Newton, 1687), Fourier analysis to understand the heat dispersion on a cannon muzzle (Fourier, 1822). We believe that the brain will be no different and require the development of genuinely new techniques suitable to capture the essential properties of neural data.

Therefore, NDS is inherently interdisciplinary, bringing principles from data science to bear on neural data to answer questions that are neuroscientifically relevant.

NDS is not unique in this regard, there have been attempts to frame other fields within a data science wrapper. Health data sciences for instance, which use principles from data science to solve problems in healthcare, using healthcare data.

HOW IS NEURAL DATA SCIENCE DIFFERENT FROM COMPUTATIONAL NEUROSCIENCE?

As we believe that the source of the data matters—especially in neuroscience—we find appropriate the description of our venture as "NDS." NDS differs from the late 20th century development of computational neuroscience (CN) insofar as CN is primarily concerned with using mathematical approaches to build—usually biophysically plausible—models of neural processes and to identify the computations corresponding to physiological processes (Marmarelis & Marmarelis, 1978). NDS transcends this framework as it is impartial to the methods used to uncover the structure of neural data and the predictions that can be made from it, but is not primarily concerned with biophysical plausibility.

DATA AS SEEN BY DATA SCIENTISTS VERSUS DATA SEEN BY NEURAL DATA SCIENTISTS

The concept of data is slightly, but importantly, different in these two fields. In neural science, data result (as in all other sciences) from a measurement process. It is *both* born and made at once.[※] A measurement process is a formal observation, assigning numbers to complex natural phenomena based on systematic rules (usually a theory of measurement). In contrast, data science presumes someone who just sent a picture or tweeted created data. In this sense, data science has a much broader definition of data. In addition to data resulting from measurement processes, data are any structured information, as evinced, for instance, by cell phone carrier "data plans." No measurements are made, but information is transmitted, which is what you are paying for.

WHAT IS A NEURAL DATA SCIENTIST?

A neural data scientist is a bold practitioner of NDS. Someone who both knows the methods necessary to log the data as well as the methods necessary to analyze, model, and understand it. They have domain expertise in a field of neuroscience, i.e., they know what the relevant questions and theories are, but their perspective is usually not confined to any particular subfield. In other words, the neural data scientist is a unicorn (at the time of this writing), but the purpose of this book is to jumpstart a dedicated unicorn breeding process to make them much more common.

WHY DO I NEED TO BE ABLE TO WRITE COMPUTER CODE?

Coding skills are essential to the neural data scientist—as they are for any data scientist. Thinking things through and working out math problems by hand are good exercises, but the ability to translate thoughts and problems into meaningful computer language is of the highest importance. Coding skills demonstrate your ability to not just think through issues, but implement practical solutions to theoretical questions.

As a neural data scientist, you are likely to be involved in the recording of the data. Recording data oftentimes means writing graphical user interfaces (GUIs), interacting with hardware, and understanding the computing machine itself. "Pure" data scientists have the luxury of blissful ignorance to all of this. You, the neural data scientist, do not have this luxury. You, the neural data scientist, must be a good programmer.

WHAT IS NEURAL DATA?

Neuroscience proper is concerned, as the name suggests, with the structure and behavior of neurons. Neuroscientists are people who study neurons. Neural data are data recorded from neurons.

However, neural data more broadly conceived encompasses any form of data that result from measurements involving the operation of part (or all) of the brain. So neural data in this sense can be recorded.[36] From systems as small as nanoscopic microtubules or as large as large groups (in neuroeconomics or social neuroscience).

CAN WE JUST ADD "NEURO" TO THE FRONT OF ANYTHING?

We are not the first to add neuro- to the front of a phrase, and we certainly won't be the last to do so. Terms like neuroastronomy, neurogastronomy, neurosociology, neuroglaciology, and neurohydrology are all terms that have been used in scientific publications. Commercial interests try to capitalize on neuroYoga and everyone is having a good time on Twitter lampooning these efforts, invoking neurometeorology and the like (Aramchek, 2014). But we think we have a legitimate case, so we are sticking with it.

WHY PYTHON?

For the past few decades, MATLAB has remained the dominant language for scientific computing and analysis in neuroscience. Known for its user-friendly interface and deep support, Mathworks has provided the academic and industrial community with a sophisticated software suite that allows data mongers to easily manage, process, and visualize datasets. Thus, MATLAB was adopted by the majority of neuroscience labs as their default computing language.

Yet MATLAB has struggled to keep up with the demands of the programming community at large, has yet to offer their product at a low cost to laboratories outside of academia, failed to gain traction in the burgeoning field of data science, neglected to keep up with developments in online deployment, and has encountered difficulties in making their product relevant or translatable outside of their quite sizeable niche.

These shortcomings allowed an opening for an alternative, yet similar, programming language. This language (*Python*) has now gained considerable momentum as a strong candidate for surmounting these shortcomings of MATLAB. Most obviously, it is freely distributed and is often included as a default language preinstalled on many computers, including all Apple Macs. Python is used and praised by programmers in fields seemingly distant from neuroscience: from webpages (YouTube, Dropbox, and Reddit, to name a few) and custom software for data science to video game development. It is renowned for its front-end agility and versatility. At the same time, it is conceptually easier to understand and has lower startup costs compared to languages like C or C++. These features have contributed to the rise of a network of diverse communities that together establish a distinct, growing, and robust niche for the use of Python as a programming language.

Yet Python remains a young, precocious newcomer in the field of neuroscience, compared to the established juggernaut—MATLAB. While the number of users is growing, overall, Python usage has remained comparatively low—in the neuroscience community, Python is about as commonly used as the relatively marginalized language R or the software package Igor. We think that this is largely owed to the fact that, while there is an existing array of textbooks outlining Python, none of them

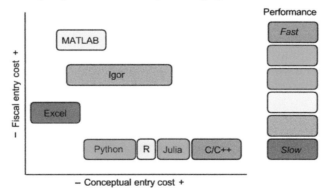

FIGURE 1.1 Features of a sample of commonly used programming languages. C/C++ is by far the fastest language, as it is lower level (and thus conceptually more difficult), compiled and closed to the machine language. MATLAB and Python are conceptually similar at first, but Python is faster, does not come with a hefty price tag, and provides for higher concept usability. Igor can be useful as an out-of-the-box program for entry-level neuroscience, but is difficult to adapt to specific needs. Excel is great for basic cut-and-paste plots, but its performance does not scale, has limited reproducibility, and a modest price tag. Julia, a relatively young language, offers performance faster than Python, but lacks the diverse packages and ecosystem that has evolved around Python.

is aimed at the neuroscience community. Though many general user issues in Python can be solved by a quick web search, systematically developed and didactically aware guides for how to process and visualize neuroscientific data are virtually nonexistent. In particular, there is no textbook on specific aspects of neuroscientific computing with Python such as data analysis, data recording, and data modeling. We hope to provide such a text with this book (Fig. 1.3).

WHY MATLAB?

MATLAB has several distinct advantages for the absolute beginner as well as for people who are not primarily interested in programming. First, there is an extensive and well-developed *help* function that is actually quite helpful. Second, it is a high-level language—most of the "plumbing" is carefully hidden from the programmer, which particularly helps beginners, as they don't get sidetracked caring about the logistics of memory handling and type casting can focus on the big picture. Third, MATLAB provides a particularly strong suite of visualization routines and functions that would take a long time to program de novo in a low-level language like C. Fourth, MATLAB is particularly optimized to represent matrices and the linear algebra operations that can be performed on them, which lends itself to handling neural data. Finally, MATLAB functions as an obvious Schelling point in neuroscience. In principle, prospective programmers in neuroscience could choose any of a large number of existing (like C or Python) or emerging (like Julia) languages. But a simple heuristic could be to pick the most popular language. This has several advantages. For one, there is a large and active community of people working on similar problems. You are thus unlikely to encounter a given problem for the first time and you can often reuse a solution that someone else already came up with. These are shared on the Mathworks "File exchange" website. In addition, other people in the lab (and even in other labs) will readily be able to read your code (at least in principle), fostering collaborations and building on an existing code base in the lab. In a way, the most obvious analogy would be the Catholic Church around the time of the Reformation. It might not be the most hip, but it has a lot of staying power and it gets, for the most part, the job done. Once one goes down the road of schisms and splintering the unity of the field, there is no reason to stop there. If one leaves MATLAB for Python, why not leave Python for Julia? Once one already has switched several times, why not go all the way and build a language that is a perfect fit for one's own idiosyncratic needs? At that point, language use would be so balkanized as to render meaningful collaboration between labs (and perhaps even within labs) impossible while at the same time taking focus away from what should matter most—developing code to answer scientific questions in neuroscience. This is a dangerous road to go down and indeed, as the history of religion illustrates, there is still only one Catholic Church, but thousands of small Protestant sects. Put differently, people switching from Python to Julia could be likened to a cancer itself getting cancer, which—until the discovery of tumor suppressor genes—was one of the most credible explanations why whales (or elephants) don't die from cancer (Nagy, Victor, & Cropper, 2007). The real reason whales don't get cancer seems to involve multiple copies of tumor suppressor genes, particularly *TP53*—whereas humans have to make do with a single copy (Abegglen et al., 2015).

WHY NOT C/C++/R/JULIA/HASKILL/JAVA/JAVASCRIPT/OCAML/PERL/PASCAL/FORTRAN/RUBY/GROOVY/SCALA/ETC.?

There are many great programming languages, each having its own tradeoff signature between particular benefits and drawbacks. At this point, Julia is a promising fourth-generation language with native support for fast parallel data processing but is the wild west's wild west, too wild in fact even for us. The language R is great for statistics and visualizations, but not particularly well suited for recording data or running experiments. If you don't already have a preexisting dataset, you are better off using MATLAB or Python because chances are your data recording routine will already use one of these languages. Yet, many hardcore machine learning aficionados and stats wonks wear knowledge of R on their sleeves. For them, it is, at once, a badge of honor and a community tag/identifier. C is great if you want to write an operating system or other speed-sensitive applications, if you have time to develop them, highlighting the typical tradeoff between runtime speed and development speed.

WHAT IS INDUSTRIAL DATA SCIENCE? HOW IS IT DIFFERENT FROM ENGINEERING?

An industrial data scientist is tasked with figuring out which algorithms are best suited to solve a company's problems. Most for-profit companies are concerned with making more money for shareholders, so data scientists in these companies work on algorithms that strive to maximize profits. This can span a range of data types and methods, from financial data models, to optimizing where advertising dollars are spent to maximize a return on investment. Data scientists can find themselves very close to particular products, on engineering teams, cleaning up messy data blobs, or generally data munging.

The neural data scientist however is conceptually closer to the industrial full-stack engineer. The full-stack engineer understands how the data are gathered, the tools used for gathering, how they are stored, how the pipeline is built and maintained, how they are visualized, how humans interact with the data/product, and how each decision about the overall flow of the data impacts the company (which typically means a good full-stack engineer can put a price tag on everything). We compare the skills of the full-stack engineer to the neural data scientist in the following table.

Full-stack Engineer	Neural Data Scientist
Understands data source	Understands the neural system in question
Builds APIs	Builds instruments for data collection
Writes GUIs for analysts	Write GUIs for technicians
Implements ETL (Extract, Transform, Load)	Writes filters to store data properly
Rapidly adopts new technologies	Rapidly tests new hypotheses
Understands business motivation	Understands scientific motivation
Builds predictive models	Builds predictive models
Maps the path to development	Maps the scientific method
Contributes to repositories	Contributes to journals

On the preceding pages, we have hoped to answer some of the most pressing questions regarding the scope of what we are trying to achieve with this book and outlining some of the issues facing data science, NDS, and neuroscience.

⌘. *Pensee on the nature of data:* We can all agree that data (*singular: datum, *not* anecdote*) need to be analyzed. But how do data come to be in the first place? Are data born or made? This harkens back to the differing notions of slavery in ancient Greece versus Rome. Greek philosophy proposed the existence of natural-born slaves, i.e., slaves whose status as a slave reflects the natural order of things, both descriptively and prescriptively (Aristotle, Politics, Book I). In contrast, Romans typically conceptualized slaves as born free but made, as a type of social status due to the vagaries of war or law that could happen to anyone, without invoking the inherent nature of a slave (Tierney, 1997). Similarly, one can wonder whether data are more akin to an apple—something that already exists in nature and just needs to be picked up or "collected" versus being more like a sonnet—not existing until human creative activity brings it into existence (see Fig. 1.2). We contend that data occupy an interesting space, as they live in the intersection of these realms, being at once born *and* made, perhaps uniquely so in the course of human events. Data are brought about by virtue of a measurement process, i.e., formal observations according to systematic rules. There are no such things as unmeasured data. Unmeasured data are not data. In this sense, the brain does not generate data unless they are observed formally. Conversely, making up data without evidentiary basis is a (perhaps the?) cardinal sin in science, as it misrepresents regions of the red space on the right as being in the purple intersection. But what are measurements fundamentally? In a sense, they are a projection. The assignment of a number or of numbers to a natural phenomenon corresponds to the projection of a usually complex and highly dimensional entity onto a relatively simple one. This is where the utility of data for science to infer the state of the natural world comes in. If this is the inherently reductive process by which data come about (it is), they are essentially shadows

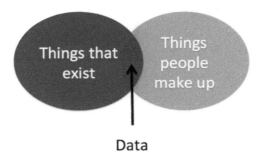

FIGURE 1.2   Data being born and made.

cast from objects in actual reality that can be used by the scientist to infer the likely nature of the object that cast it. Like any cognitive system, we cannot leave Plato's cave (see Fig. 1.3), but we can look at the shadows to infer what the likely object was, in a probabilistic fashion. It is important to keep in mind that data are not themselves reality and neither are the models of reality that we create on their basis. It is important not to confuse shadows with objects—the same object can give rise to many different kinds of data, depending on the measurement process. This has implications for what this process of measurement should be called. Saying that data are created or generated emphasizes the human creativity that brings them about and seems too arbitrary. Saying that data are collected, acquired, obtained, or gathered emphasizes the fact that we are trying to quantify relevant aspects of a preexisting natural phenomenon, giving no credit to the human creativity involved in the measurement process. Saying that data are measured is an awkward truism. The Latin root of data, "a given thing" does not help us. If anything, data are ironically named, as they precisely *not* given things. There is nothing given about them unless a measurement makes it so. What was meant with the latin term "data" when it was introduced in the 17th century was "the givens", the facts as they are, as the basis from which we reason. But epistemology has progressed a great deal since then. We no longer accept data as "given". To emphasize the unique status of data among human affairs, perhaps it would be best to propose a neologism, "data are being marned" (being both born and made). Recognizing that this probably won't catch on, this leaves us with the (somewhat bland but perhaps most precise of the available options) data being "logged" or "recorded," as it also implies that someone else could record something else—the original object is still there. In this sense, data recording is analogous to a tree falling in a random forest. There in one natural process—the falling tree creating a pressure wave in air. But different observers can record different data from this process. At heart, data recording is a one-to-many mapping, in contrast to collecting, which implies a one-to-one mapping. Another implication of this choice of term is that there can be no such thing as "simulated data," which is perhaps better called *sata* (Wallisch, 2014). All of this is not idle semantics. Fuzziness of language usually implies fuzziness of thought and can lead to confused action. Being as precise as possible in one's use of language pays ample dividends later on, by helping to avoid avoidable confusion. Finally, a theorist *can* collect data—from experimentalists. But at some point, someone needs to make measurements (of a natural process) in order to record data in the first place.

FIGURE 1.3   Plato's cave.

CHAPTER

2

From 0 to 0.01

WHAT IS THE GOAL OF THIS CHAPTER?

The goal of this book is to take people from 0 to 1.[*] The goal of this chapter is to take people from 0 to 0.01. To get there, we only presume that you can *read*, *write*, and do *arithmetic* (the 3 R's). Here, we will provide the basic tools to install all of the software you will need so you can start writing computer code and everything else that you need to get going. A deeper understanding of statistics, machine learning, and neuroscience will come about from writing your own programs, running them, and looking at their outputs—often quite literally, as they will usually produce figures.

HOW DO I GET STARTED CODING?

To install Python, we endorse the Anaconda distribution. Anaconda is a free download of Python that includes over 400 common and useful libraries. In this chapter, we'll be using the iPython shell, which we use interchangeably with the term Python. To run iPython, open *terminal* (on Macs), type *ipython*, and press return in the terminal or command window.

As this book went to press, Anaconda can be downloaded at: https://www.continuum.io/downloads.

If this link is broken, you can search online for the keywords *Python* and *Anaconda*—the download link should be on the first page that comes up. This book was written for the purpose of teaching Python 2.7, so you will get the most use out of this book if you also use Python 2.7. The next stable release, Python 3.5 is great, but not all external packages support it yet and most Python programmers still use Python 2.7 at the time of this writing, so we're also sticking with Python 2.7 (also called Python 2), which we'll just refer to as Python from here on out. If you're reading this book at a

Neural Data Science.
DOI: http://dx.doi.org/10.1016/B978-0-12-804043-0.00002-7

later time, and are likely using a later version of Python, most things—importantly all concepts—will transfer although there might be slight syntactical differences that are documented on several websites that come up if you search online for "python 2 vs 3." For some examples of key differences, see Appendix C.

As for MATLAB, simply buy and install it. It should be self-explanatory. For the purposes of this book, we use release 2016a. Mathworks is committed to new MATLAB releases twice a year, one in the spring and one in the fall. As for Python, all concepts will transfer if you use this book with a later version of MATLAB, but do expect there to be slight syntactical differences that accumulate over time. Like with Python, we will document some key differences in Appendix C.

WHAT'S THE COMMAND LINE? WHAT'S THE ENVIRONMENT?

The command line and environment are like your pencil and notebook. This is where you enter commands that tell the computer what to do—interpreted via Python or MATLAB (POM, hereafter). To do the basic command line exercises here in Python, open the program *terminal* and type *ipython* and hit the return key. This will put you into an interactive Python environment. You can also use the program Spyder that comes with the Anaconda distribution, it is quite similar in feel to MATLAB. Programming aficionados will often use a range of text editors depending on what they are trying to accomplish; for more discussion on how to use different text editors, see Appendix C.

HOW ARE PYTHON AND MATLAB DIFFERENT?

Because we are trying to get to the smallest possible incremental increase in proficiency level in this chapter, we can only begin to address this issue here, by showing fundamental but practical differences. If you want to dig deeper, there is an entire appendix dedicated to this, as there has to be (Appendix A).

HOW DO I DISPLAY SOMETHING ON THE SCREEN?

Python	MATLAB
```>>> print 'Hello World'``` ```Hello World```	```>> disp ('Hello World')``` ```Hello World```

HOW DO I DO ARITHMETIC IN PYTHON OR MATLAB?

All arithmetic (addition, subtraction, multiplication, and division of single numbers) works just like you would expect. Simply type in the command line what you would into a calculator:

Python	MATLAB

`>>> 5+7`
`12`
	`>> 5+7`
	`ans =`
	`12`

Note that the MATLAB output adds an "ans = " line before giving you the output. This indicates that the output was assigned to a temporary variable. We will worry about this later (actually, we won't worry, but we will explain it in detail). For now, understand that for the sake of consistency, matching lines of MATLAB and Python input/output line by line, we will not print the `ans` output. Do not be startled or surprised if it does show up on your screen. We predict that it will.

Subtraction works the same way as addition, using the dash or minus sign (–):

Python	MATLAB
`>>> 5-7` `12`	`>> 5-7` `12`

Multiplication is implemented by using the asterisk sign (*):

Python	MATLAB
`>>> 5*7` `35`	`>> 5*7` `35`

However, division works seemingly different between Python and MATLAB:

Python	MATLAB
`>>> 7/5` `1`	`>> 7/5` `1.4`

MATLAB gets it right here, as 7 divided by 5 is not 1, but 1.4. What Python is doing here is called *integer division*. Integer division is dividing one integer by another integer and results in the output of another integer. Integers are whole, natural numbers such as 1, 4, or 17, but not 2.5. As the result of dividing 7 by 5 is fractional, the result of integer division is the integral part of the result, ignoring the fractional remainder. In effect, this corresponds to rounding, specifically to rounding *down*:

Python	MATLAB
`>>> 9/5`	`>> 9/5`
`1`	`1.8`

This introduces our first example of data types in Python. If you perform operations on integers, Python assumes that it should also result in an integer. Python lets you write out equations without specifying data types, and tries to infer later what kind of numbers it's dealing with. In computer science, this is called *weak typing*. Many other programming languages (such as C) require that you explicitly specify data types (which is both more precise and verbose, and called *strong typing*). So what if we want to do operations on decimal numbers that result in decimal numbers, not integers? We can specify this in one of two ways. One of them is to put decimal zeros after each number. The presence of a decimal point signifies to Python that you care about what happens after the decimal point in terms of the result.

Python	MATLAB
`>>> 7.0/5.0`	`>> 7.0/5.0`
`1.4`	`1.4`

As you can see, the MATLAB output is not modified by adding the decimal point. This is because the default data type (unless specified otherwise) in MATLAB is a floating-point number with double precision. Floating point data types represent real numbers with a certain number of significant digits. Double precision means that the number of significant digits is large. A single number represented in double precision floating-point format has 64 bits, or 8 bytes, whereas integers take up only 8 bits or 1 byte in memory. So double precision floating point numbers take up a lot of space in memory but are able to represent numerical results with sufficient precision for (almost) all practical purposes in science.

Another way to make sure that Python treats numbers as floating-point numbers is to specify this explicitly, which is also the first time we introduce the notion of a function. A function is a program (built-in, self-made, or made by another user) that contains a set of instructions, takes a number of inputs and returns an output. Which inputs the function takes is specified by parentheses after the function name in POM. The parentheses signify that what is between them should be interpreted (and treated) as inputs to the function that was invoked. The function name to convert numbers to double-precision floating-point numbers in Python is `float`, whereas it is `double` in MATLAB:

Python	MATLAB
`>>> float(7)/float(5)`	`>> double(7)/double(5)`
`1.4`	`1.4`

As you can see, the `double` is silent in MATLAB, as it is (for historical reasons, double precision being the engineering industry standard) the default data type in MATLAB and doesn't need to be spelled out.

MATLAB differentiates several types of floating point number, "single precision" and "double precision." As already stated, double precision numbers consist of 64 binary digits or bits and take up 8 bytes in memory, whereas single precision numbers consist of 32 binary digits and take up 4 bytes in memory apiece. So single precision saves space, but can be too imprecise for scientific applications if they involve more than seven significant figures. Consider how the real numbers 1/7 and 1/13 are represented in double versus single precision. In MATLAB, the `single` function species that we would like represents its inputs as a single precision floating point number:

Python	MATLAB
`>>>1./7.`	`>> 1/7`
`0.14285714285714285`	`0.142857142857143`
	`>> single(1/7)`
	`0.1428571`
	`>> 1/13`
	`0.076923076923077`
	`>> single(1/13)`
	`0.0769231`

In single precision, 1/7 is correctly rounded down at 7 after-decimal-digits whereas 1/13 is correctly rounded up at 7 after-decimal-digits.

So far, we have only discussed how numbers are represented internally, not how they are displayed. In addition to (and independently of) how numbers are represented, which affects the precision of computations done on them, you can also specify how they are displayed, which does not. To do so, we use the `format` function in MATLAB and `decimal` in Python. `format long` displays 15 digits after the decimal point for doubles and 7 for singles, `format short` displays 4 digits after the decimal point and `format bank` displays 2 digits after the decimal point. In Python, specifying precision is much more complicated but also much more versatile. In the code below, we show the Python code that will replicate the corresponding MATLAB behavior, but if you want to truly understand it, we refer you to Appendix A, where we will explain this detail.

To wit:

Python	MATLAB
| ```
>>> from decimal import *
>>> getcontext().prec = 15
>>> Decimal(1) / Decimal(7)
0.14285714285714285
>>> getcontext().prec = 4
>>> Decimal(1) / Decimal(7)
0.1429
>>> getcontext().prec = 2
>>> Decimal(1) / Decimal(7)
0.14
``` | ```
>> format long
>> 1/7
   0.142857142857143
>> single(1/7)
   0.1428571
>> format short
>> 1/7
   0.1429
>> single(1/7)
   0.1429
>> format bank
>> 1/7
         0.14
>> single(1/7)
         0.14
``` |

The latter is generally used to represent currencies; there are other formats as well. If you are unsure what formats are available and how to specify them, you can always invoke the `help` function, in MATLAB: *help format*, and from the iPython shell: `help(format)`. More on the help function shortly.

We recognize that the topic of numbers, how they are represented internally and displayed to the user, can appear a bit dry and obscure to the casual reader. However, in our experience it is good to establish a preliminary understanding of these matters early on to avoid confusion as to why POM is doing what it is doing. Recall how we got into this discussion—by dividing 7 by 5 in Python and getting the puzzling result of 1. With that in mind, this is all we want to say about numbers, how they are represented and displayed in this book. If you are looking for a more principled in-depth treatment of this topic and of numerical representations of real numbers in memory, see Chapter 4.4.4 of Wallisch (2014).

Going back to integers, we can replicate Python's default behavior in MATLAB by explicitly stating that our number is represented by an integer. Just like for floating point numbers, MATLAB does differentiate several types of integer.

The smallest (in memory) can represent 256 states, and takes up 1 byte (8 bits in memory). Conversely, the `int` is silent in Python.

Python	MATLAB
`>>>7/5`	`>> int8(7)/int8(5)`
`1`	`1`
`>>>int(7)/int(5)`	
`1`	

HOW DO I INPUT EXPONENTS IN PYTHON AND MATLAB?

Exponents in Python are implemented with a double asterisk (`**`) and with the carrot (`^`) in MATLAB, so five squared (5^2), five to the sixth power (5^6), and the square root of 5 ($5^{1/2}$ or $5^{0.5}$) is:

Python	MATLAB
`>>> 5**2`	`>> 5^2`
`25`	`25`
`>>>5**6`	`>> 5^6`
`15625`	`15625`
`>>>5**0.5`	`>> 5^0.5`
`2.2360679774997898`	`2.23606797749979`

WHAT IS THE ROLE OF BLANK SPACE IN WRITING CODE, IF ANY?

Blank spaces are of critical importance in Python when they occur at the front of the line (we'll talk about this much more later, when we tackle `for` loops), but are ignored between numbers and operators in both POM:

```
Python                                          MATLAB

>>> 5 ** 2                                      >> 5^2
25                                              25
>>>5        **6                                 >> 5^6
15625                                           15625
>>>5**           0.5                            >> 5^0.5
2.2360679774997898                              2.23606797749979
```

This flexibility in spaces can result in unreadable code, so it's good to pick a style and stick with it; e.g., using a single space in between each character is acceptable and will result in clean-looking code. How you get there, i.e. by using tabs vs. spaces is a matter of great controversy, trading off precision vs. efficiency.

WHAT IS THE ORDER OF OPERATIONS IN PYTHON AND MATLAB?

Both languages follow classical mathematical order of operations: parentheses are evaluated first, exponents are evaluated next, then multiplication and division, and addition and subtraction are evaluated last. If operations on either side of a value are equal in order, the equation is evaluated from left to right. Note that if you add an *int* to a *float*, the result is a *float* in Python, but an *int* in MATLAB:

```
Python                                          MATLAB

>>> (5 + 4.0/2.0)**2                            >> (5 + 4/2)^2
49.0                                            49
>>> (5*4.0/2.0)**2 + 5                          >> (5*4/2)^2 + 5
105.0                                           105
>>>5.0*4.0/2.0**2.0 + 5                         >> 5*4.0/2.0^2 + 5
10.0                                            10
>>>5*4/2**2+5                                   >> 5*4/2^2 + 5
10                                              10
```

WHAT ARE FUNCTIONS?

We already encountered some functions above, in the form of help, double, int8, and format. Generally speaking, a function is a command that makes something happen in POM, e.g., change the way all numbers are displayed in POM.

However, most of the time, functions are operators that take inputs and transform them to outputs in a systematic fashion. For instance, the corresponding `rand` functions in Python and MATLAB take as an input how many random numbers you want to create, and returns that many random numbers. Say you want to create five random numbers drawn from a uniform distribution between 0 and 1 that are output as a horizontal vector. The appropriate code for POM is:

Python	MATLAB
```>>> import numpy```	```>> rand(1,5)```
```>>> numpy.random.rand(5)```	0.757740130578333
```array([0.31075661, 0.45360828,```	0.743132468124916
```0.95271364, 0.14573102,```	0.392227019534168
```0.63847612])```	0.655477890177557
	0.171186687811562

Do not be alarmed if the numbers returned by your invocation of the random number generation function do not produce these exact numbers. That's the point, they are supposed to be random. Note that even if you invoke this function repeatedly, you (hopefully) will get different outputs each time. We'll revisit the random number generator in a more principled fashion later, where we explain how to control the generation of a pseudorandom number stream more precisely.❀❀

You can even "nest" functions, where POM applies two functions in one line (one after the other) starting with the "innermost" function. In this example, we take the square root of a negative number *after* taking its absolute value first.

Python	MATLAB
```>>> numpy.sqrt(abs(-4))```	```>> sqrt(abs(-4))```
2	2

In a later chapter, we'll show you how to write your own functions if those that already exist in POM should be insufficient.

WHAT ARE PYTHON PACKAGES? WHAT ARE MATLAB TOOLBOXES? ARE THESE DIFFERENT FROM LIBRARIES?

You can save the functions you write. If you have written enough of them on a particular topic, you can call it a "toolbox." In fact, some people have already done this and made them publicly available. Numpy is a great example of a publically available Python package that we used above to create the random numbers in Python.

Packages in Python are the toolboxes that you'll use for coding. There are many parts of Python that are built-in, meaning that you don't need to install or load any extra packages. Some examples of these built-in functions in Python include sum, which calculates the sum of elements, and abs, which returns the absolute value of the input. In Python, you need to explicitly load packages before you can use the functions they contain. For example, we'll be using the package numpy frequently in this book. To load numpy (or other packages), we use the import function, which is why we typed import numpy above, before we could use its rand function. Numpy, like many Python packages, groups functions further. For instance, its randomization functions are contained in a subpackage called random. One of the functions contained in random is the rand function, which draws from a uniform distribution. There are others. Then specify which exact function you want, you use a period. This is also called accessing the "methods" of a package or subpackage. As seen above, to invoke numpy's rand function, we type numpy.random.rand() and put the inputs within the parentheses. Once you import a package, its contents are available for the duration of your coding session, i.e., you do not have to import numpy each time you want to use function in numpy.

Just like Python packages are special purpose libraries of functions, MATLAB has dedicated toolboxes of advanced functions that have to be bought separately and are not included in the standalone version of MATLAB. MATLAB toolboxes used by the Neural Data Scientist are likely to include the Statistics and Machine Learning Toolbox, the Signal Processing Toolbox and the Neural Network toolbox, among others.

HOW DO I GET HELP?

We have already called for help before, when trying to see what argument the format function takes. MATLAB comes with over 7000 functions at this point. They all have their idiosyncratic syntax—that also sometimes changes with new versions of MATLAB. How are you supposed to know how to use a function or whether it even exists? The answer lies in help, which is itself a function. If you type help(functionName), you will get an inline answer as to what the function is supposed to do, what inputs (arguments) it expects, how it works and (if you are lucky) an example. There is no shame in using the help function (even if it can be hard to ask for help). What makes MATLAB one of the most user-friendly programming languages for beginners is that the help is actually helpful. A lot of introductory MATLAB books are nothing but annotated help files. Another way to get more in-depth help in MATLAB is to use the function *doc*. For instance, doc rand will bring up a more elaborate help page about the rand function than *help rand* would. Finally, the function browser hovers near the cursor and looks like f_x—if you click on it, you can search the help file for all functions that might be helpful. Use it.

IPython has a similar help function, though lacks the *doc* functionality. Type help(functionName) to view the documentation (press q to exit the help function). In addition, a large community of Python programmers provide help for errors on the website *stackoverflow.com*, and also provides documentation for all common packages at https://docs. python.org/2.7/.

The most common reason to call for help is when encountering errors. Perhaps the best way to figure out why you are getting an error is to paste your exact error message into Google (this works well for POM). Odds are the top link will be to a stackoverflow or Mathworks page, but oftentimes other support websites and communities will have figured out the error and provided good examples for how to fix your code.

WHAT ARE VARIABLES?

In POM, we usually don't just want to calculate something on the spot, then forget about it. We want to retain the results of our labor for later use. The way to do so is to reserve a spot in memory and assign the outcome of our calculations to it. Importantly (and somewhat confusingly, if you are used to look at a lot of equations), the assignment operator in POM is =.

This does not mean that this is a test for equality or that equality is implied.[※※※] The assignment operator means "assign whatever is on the right of it to what is on the left." That's it. Conversely, a test for equality is indicated by == in POM. In the code below, we first assign a value of 5 to a spot in memory we call "A" and a value of 2.5 to another spot we call "B" (A and B are called variables). We can test whether the value of A equals the value of A (true), A equals B (false) and A equals C (true). C is defined as 2*B.

Python	MATLAB
`>>> A = 5`	`>> A = 5`
`>>> B = 2.5`	`>> B = 2.5`
`>>> C = 2*B`	`>> C = 2*B`
`>>> A == A`	`>> A == A`
`True`	`1`
`>>> A == B`	`>> A == B`
`False`	`0`
`>>> A == C`	`>> A == C`
`True`	`1`

And he created a matrix C and saw that it was good. He also noted that the syntax is very similar in POM, and there was great rejoicing in all the corners of the earth. Although Matlab uses conventions of Boolean logic to represent the truth values of the output ("1" for true and "0" for false), whereas Python is more explicit.

HOW CAN I ACCESS OR DISPLAY WHAT IS CONTAINED IN A GIVEN VARIABLE?

Simply by typing its name. We can now demonstrate that format doesn't change anything about how a number is represented or the contents of a variable, only how the contents of a variable are displayed. In MATLAB we can specify `format` to be `long`, or `bank`. In the ipython shell, we can specify the `precision`, or how many decimal places to show, using the magic operator `%precision XX`, where XX is the number of decimal places to display. We'll explain magic commands in ipython at the end of the chapter.

Python	MATLAB
>>> %precision 15	>> format long
>>> A = 1./7.	>> A = 1/7
>>> A	A =
0.142857142857143	0.142857142857143
>>> %precision 2	>> format bank
>>> A	>> A
0.14	A =
>>> %precision 15	0.14
>>> A	>> format long
0.142857142857143	>> A
	A =
	0.142857142857143

WHAT IS "ANS" IN MATLAB?

If you don't specify a variable to which to assign the results of your computations, MATLAB automatically creates a temporary variable "ans" for you and assigns the result of the computation you just performed to it. "ans" is updated every time you do this, so it is not a good place for the long-term storage of data. You can use it for successive computations, like in the example below. In the Python interactive shell, the underscore or "_" is updated every time as the last answer.

Python	MATLAB

``
>>> 5 + 5
10
>>>_ + 5
15
>>>_ +_
30
``
	``
	>> 5 + 5
	ans =
	    10
	>> ans + 5
	ans =
	    15
	>> ans + ans
	ans =
	    30
	``

Note that the underscore in Python will not necessarily update within scripts; rather, the underscore is a quick method for the last variable only in the interactive shell.

WHAT CAN WE CALL OUR VARIABLES?

Variables are named slots in memory. By invoking their name, we can access what is stored at that point in memory. This brings up the question of what you can, or should, call your variables. We recommend naming them something descriptive and indeed recommend to use what is sometimes called "camel case." For instance, if you want to store the number of participants in a study in a variable "A," you can do so, but you won't remember this in 6 months when you need to reanalyze the data because reviewer 2 asked for it. Instead, we suggest to call it something like numParticipants. This is an example of camel case. We could have used underscores, but they are unwieldy, e.g., num_participants. The capital letters (like the humps of a camel) neatly parse the stream of characters. With this in mind, you can call your variables whatever you want, but there are some names that are illegal (e.g., those that start with a number or have a space in it) and some that are strongly discouraged. We strongly discourage using a variable name that is already the name of a preexisting function. For instance, we already used the abs function. We can assign a variable with that name, but can no longer use the abs function like you used to after doing so, e.g.:

Python	MATLAB

``
>>> abs(-2)
2
>>> abs = abs(-4)
>>> abs(-2)
--------------------------------- 	
TypeError Traceback (most
recent call last)
<ipython-input-134-2f17e538472a>
in <module>()
----> 1 abs(-2)
 	
TypeError: 'int' object is not
callable
``
	``
	>> abs(-2)
	2
	
	>> abs = abs(-4)
	abs =
	    4
	
	>> abs(-2)
	
	Index exceeds matrix dimensions.
	``

POM throws an error; in MATLAB in the color of blood to get your attention (unless you specify it otherwise in the preferences). See the "frequently made mistakes" appendix to understand in detail what is going on. For now, just keep in mind not to name your variables like a preexisting function if you are still planning on using that function in the future.

If you have done so accidentally, you can delete the variable from memory (and use abs again, in this instance) by typing the code below.

Python	MATLAB
| ```
>>> del(abs)
>>> abs(-2)
2
``` | ```
>> clear abs
>> abs(-2)
2
``` |

The `clear` function deletes variables from working memory (in MATLAB, this is called the "workspace"). If you type `clear` or (for older version of MATLAB) `clear all`, all variables are deleted from memory. If you want to specify to remove a single variable while leaving the others untouched, simply provide the name of the variable that you want to delete as an argument, like in the example above. This works fine, unless you accidentally call a variable `clear`. Do *not* do this, as it is harder to recover from, like if there is a fire in the fire department:

Python	MATLAB
| ```
>>> abs = 3
>>> del = 5
File "<ipython-input-30-b1b9fc5a7c5f>", line 1
    del=5
        ^
SyntaxError: invalid syntax
``` | ```
>> abs = 3
abs =
    3
>> clear = 5
clear =
    5
>> clear abs
Error: "clear" was previously used as a variable, conflicting
with its use here as the name of a function or command.
See "How MATLAB Recognizes Command Syntax" in the MATLAB
documentation for details.
>> clear clear
Error: "clear" was previously used as a variable, conflicting
with its use here as the name of a function or command.
See "How MATLAB Recognizes Command Syntax" in the MATLAB
documentation for details.
>> clear
clear =
    5
help clear
clear is a variable of type double.
``` |

If you are unsure if a potential variable name—lets' call it `variableName`—corresponds to an existing function, you can type `which variableName`. If it already exists, it will show you the path of where it exists, if it does not, it will tell you that it does not. In Python, you are simply not allowed to use the variable named `del`. As an aside, if you want to clear all variables, use the magic command `%reset` (more on magic commands later).

WHAT IS A VECTOR? HOW DO I STORE A VECTOR IN POM?

So far, we have only assigned single numbers (or "scalars") to variables. We have implicitly created a vector when creating the five random numbers above. We can also assign a collection of numbers to a variable at once. Such a collection has a substructure—if it is a vector these numbers can be arranged either in rows or columns. So a vector is a "line" or stack of numbers and there are two types: Horizontal and vertical lines.

a. **"Row vector" A with 5 elements (number of rows = 1, number of columns = 5)**

Python	MATLAB
`>>> A = numpy.array([1, 2, 3, 4, 2])`	`>> A = [1, 2, 3, 4, 2]`

Note that the notation is essentially the same in POM. However, in MATLAB, the comma (",") is optional. A vector where the comma is missing assumes that a comma was separating the numbers. In Python, forgetting to spell out the commas will throw an error. Also, note that in Python, vectors are a special type of "array" (where one of the dimensions is 1), which is why we specify to create an array in the example above. More on this shortly.

b. **"Column vector" B with 3 elements (number of rows = 3, number of columns = 1)**

Python	MATLAB
`>>> B = numpy.array([[3],[5],[4]])`	`>> B = [3; 5; 4]`

Note that we just introduced a new kind of parentheses, the square bracket [ ]. Whereas inputs to functions or arguments are denoted by regular parentheses ( ), collections (or sets) of numbers are denoted by square brackets in POM. They are *not* interchangeable. If you open a parenthesis, it has to be closed by a matching parenthesis. If you open a square bracket, it has to be closed by a matching square bracket. Mixing and matching will throw errors. Disregarding this is one of the most frequent sources of frustration for the beginner.

c. **Determining the dimensionality of these vectors**

Each variable has a dimensionality. Dimensionality refers to the number of rows and columns (and sheets, workbooks, etc., if there are more dimensions) the variable contains. For now, we are dealing with vectors, so one of these numbers will be 1 (in Python, 1-dimensional row vectors implicitly don't have a second dimensional length). To assess the dimensionality of a given variable, we invoke the `size` function in MATLAB, and the `numpy.shape` function in Python:[a]

Python	MATLAB
`>>> numpy.shape(A)`	`>> size(A)`
`(5,)`	`ans =`
`>>> numpy.shape(B)`	`        1      5`
`(3,1)`	`>> size(B)`
	`ans =`
	`        3      1`

d. Converting row to column vectors and vice versa

There are multiple ways to convert row vectors to column vectors and vice versa. The technical term for this is to "transpose" a vector. Here is how you do this in POM:

Python	MATLAB
`>>> B.T`	`>> transpose(B)`
`array([[3, 5, 4]])`	`ans =`
`>>> A.T`	`        3      4      5`
`array([1, 2, 3, 4, 2])`	`>> A'`
	`ans =`
	`        1`
	`        2`
	`        3`
	`        4`
	`        2`

As you can see, in MATLAB, there are two functions: `transpose` and a single straight scare quote ' that accomplishes this. These two ways are equivalent, but the single square quote way saves some typing.

[a]This highlights a fundamental difference between Python and MATLAB. MATLAB's native data type is a matrix, so it is natural to transpose it, even if one of the dimensions of the matrix is 1 (a vector). Python does not treat 1d arrays as either rows or vectors, and thus will only return the length in one dimension. Python doesn't assume 1d arrays to be rows, and therefore doesn't know what transpose means to a 1d array. The shape of a 1d Python array with 5 elements is thus (5,). If it helps you to think of this as a 5 by 1 matrix then that is great, just know that there is not an explicit dimension about which you can transpose it.

HOW DO I CALCULATE THE SUM AND MEAN OF ALL VALUES IN A VECTOR?

What can we do with this? Most functions can operate on an entire collection of numbers at once, not just on scalars. For instance, we can sum all of them up with the sum function. Or take the average with the mean function.

Python	MATLAB

```>>> a = numpy.array([1,2,3,4])```
```>>> numpy.sum(a)```
```10```
```>>> sum(a)```
```10```
```>>> numpy.mean(a)```
```2.5```
	```>> A = [1 2 3 4]```
	```A =```
	``` 1 2 3 4```
	```>> sum(A)```
	```ans =```
	``` 10```
	```>> mean(A)```
	```ans =```
	```2.5```

WE NEED TO TALK ABOUT THE ECHO

In MATLAB, the "echo" is on by default. This means that unless we suppress it, we will see the output of a given operation. Sometimes, this is what you want—as a manipulation check to see what you did is what you want and that it is good. However, sometimes (say when creating a million random numbers and allocating it to a variable for later use) you don't want to see the output. So we need to find a way to suppress the output, but still do the command. The most straightforward way to do this in MATLAB is the semicolon operator at the end of a line. Note that this does not mean "execute." Pressing enter executes a line. The semicolon simply suppresses the output. But do also note that the semicolon operator is overloaded in three different ways in MATLAB. This is bad from the perspective of usability, but it is what it is. We already encountered it in the form of "vertical concatenation" above, when declaring the column vector B.

a. Using semicolon to suppress the echo. Note: One e 6 is scientific notation for a million

Python	MATLAB

```>>> A = numpy.random.rand(1e6)```
```>>> sum(A);```
```>>> sum(A)```
```499781.53261511476```
	```>> A = rand(1e6,1);```
	```>> sum(A)```
	```ans =```
	```500321.24992532```

Note that if you forgot to put the semicolon at the end of the line in the MATLAB column, it might be of interest to know that Ctrl-C stops the execution of POM.

b. Using semicolons to separate rows of a column vector C, to suppress the echo and to concatenate multiple statements in one line:

Python	MATLAB

`>>> C = numpy.array([1, 2, 3]) D = sum(C); E = D * numpy.mean(C)`
 `>>> E`
 `12.0`
	`>> C = [1; 2; 3]; D = sum(C);`
	 `E=D*mean(C)`
	 `E =`
	    `12`

Note that in Python, having an assignment operator (=) in the mix autosuppressed the echo.

HOW DO I CALCULATE THE LENGTH OF A VECTOR?

Python	MATLAB

`>>> C = numpy.array([1,2,3])`
 `>>> len(C)`
 `3`
 `>>> numpy.max(numpy.shape(C))`
 `3`
	`>> length(C)`
	 `ans =`
	    `3`
	 `>> max(size(C))`
	 `ans =`
	    `3`

Note that this is as good a place as any to talk about the fact that there are usually *many* ways to accomplish the same thing. For instance the function `len` (in Python), `length` (in MATLAB) returns the largest dimension of the array. As we are dealing with vectors, this will be the dimension that is not 1. But you can achieve the same by nesting the functions `size` (which returns a vector of dimensionalities) and `max` (which returns the largest value of a vector). We like to think of (scientific) programming as creative, but a kind of creativity that is useful. Relish in the possibilities to solve problems in a creative fashion. There are usually substantial programmatic degrees of freedom, trading off readability, execution speed, robustness, and conciseness of code when trying to accomplish any given programming goal.

But we should move on. The next natural extension of this unfolding progression is to wonder what happens if you have an array with numbers arranged in multiple dimensions.

WHAT ARE MATRICES, WHAT ARE ARRAYS?

We have already encountered a special type of array, namely vectors—those are arrays where one of the dimensions is 1 and where there are only two dimensions (rows and columns) total. Vectors are stacks of single numbers. Arrays more generally are stacks of vectors. Importantly, the number of elements in each row and column have to be equal, there can be no holes. To repeat, the "dimensionality" of an array is the number of elements in a row or column for a given array.

In MATLAB, these arrays are called "Matrices." It is the native data type of MATLAB. MATLAB itself stands for "matrix laboratory"—an environment created for the efficient manipulation of matrices. MATLAB is optimized for matrix operations, such as those used in linear algebra. The equivalent data type in Python is called a *numpy array* (similar to the Python *list*, which we address later). The *numpy array* can be multidimensional, and is not to be confused with the *numpy matrix*, which is a strictly two-dimensional array.

What distinguishes MATLAB matrices from vectors is that a vector is a special case of a matrix where one of the dimensions is 1. But even scalars are represented as matrices in MATLAB, they are simply 1 × 1 matrices. Let's play around with matrices and arrays. Note that we will use the terms "matrix" and "array" interchangeably from here on out. If we say "array," it means "matrix" in MATLAB and if we say "matrix" it means "array" in Python.

a. Declaring matrices

Python	MATLAB

```>>> A = numpy.array([[1, 2, 3],[4,5,6],[7,8,9]])```
```>>> numpy.shape(A)```
```(3, 3)```
```>>> numpy.size(A)```
```9```
	```>> A = [1 2 3; 4 5 6; 7 8 9];```
	```>> size(A)```
	``` 3 3```
	```>> numel(A)```
	``` 9```

In the first line, we declare a 3 × 3 matrix with nine total elements. We then assess that this is so with the suitable functions. Note that they are different in MATLAB and Python—the function `numpy.size` in Python corresponds to the function `numel` in MATLAB and the Python function `numpy.shape` corresponds to the `size` function in MATLAB.

b. Operating on matrices and matrix indices

Let's do some matrix surgery. We now access and return the elements in the third row. We then replace the number in the second column in the row with 55.

Python	MATLAB
```\n>>> A[2,:]\narray([7, 8, 9])\n>>> A[2,1] = 55\n>>> A[:2,1:]\narray([[2, 3],\n [5, 6]])\n```	```\n>> A(3,:)\n 7 8 9\n>>A(3,2) = 55;\n>>A(3,:)\n 7 55 9\n>> A(1:2,2:end)\nans =\n 2 3\n 5 6\n```

The colon operator means "elements from to" in POM. If no start and end element is indicated, it means "all elements." The last line means "rows 1:2 and columns 2:end," where end refers to the last column of the matrix in MATLAB.

Note that to access the third row in MATLAB, we simply typed "3." The corresponding index number in Python is 2. Why?

The reason for this is that MATLAB is indexed to 1, meaning that the first element of a matrix is element number 1. Trying to access the 0th element throws an error. Conversely, the first element of a Python array is 0.

As you can see, this matters. To access the *same* elements of a matrix when translating from MATLAB to Python, you have to subtract 1 or add 1 to the indices if you go in the other direction.

More on this in Appendix A, where we compare MATLAB and Python more systematically.

BACK TO VECTORS: HOW TO VECTORIZE A MATRIX?

An array typically contains multiple columns and rows. There are many reasons why one might convert its contents into a single vector, such as when you want to perform an operation on all elements of a matrix, but don't want to do nested loops, which are inefficient (more on this later when we talk about loops). This process is also called "linearization." There many ways to do it in either language. The two most common ones are explained here. The idea is that we take our array A with the 3×3 structure and "flatten" it into a 9×1 or a 1×9 array. This is a lossless conversion—we don't lose any of the numbers in the array, we simply rearrange them.

Python	MATLAB
```\n>>> A.flatten(0)\narray([1, 2, 3, 4, 5, 6, 7, 55, 9])\n>>> A.flatten(1)\narray([1, 4, 7, 2, 5, 55, 3, 6, 9])\n```	```\n>> A(:)'\n 1 4 7 2 5\n 55 3 6 9\n>>reshape(A,[1,numel(A)])\n 1 4 7 2 5\n 55 3 6 9\n```

There are several things we want to note about this. First, there really are several ways to do this in either language—the outcomes are the same. Second, the colon operator ":" can have multiple purposes. It doesn't just mean "all elements in a row or column," as we have seen before, it also means vectorize if it is given to a matrix without any other inputs. The transpose is not strictly necessary on the MATLAB side, but it creates a row vector to match the Python output. Third, the default MATLAB way of vectorizing is column-wise. As a matter of fact, almost all operations in MATLAB are column-wise operations. For instance, taking the mean (without further arguments) is the column-wise mean. As is the sum. If you want to change that, you need to transpose first. As seen in the following table.

Python	MATLAB
```>>> B = A.T```	```>>B = A';```
```>>> numpy.mean(A,0)```	```>> mean(A)```
```array([4., 20.66666667, 6.])```	``` 4 20.6666666666667 6```
```>>> sum(A)```	```>> sum(A)```
```array([12, 62, 18])```	``` 12 62 18```
```>>> B.flatten(1)```	```>> reshape(B,[1,numel(B)])```
```array([1, 2, 3, 4, 5, 6, 7, 55, 9])```	``` 1 2 3 4 5```
```>>>numpy.mean(B,0)```	```6 7 55 9```
```array([2. , 5. , 23.66666667])```	```>> mean(B)```
```>>> sum(B)```	``` 2 5 23.6666666666667```
```array([6, 15, 71])```	```>> sum(B)```
	``` 6 15 71```

To make this even more clear, consider the following matrix and the operations we do on it:

Python	MATLAB
```>>> B = numpy.array([[1,2,3],[4,5,6],[7,8,9]])```	```>> B = [1 2 3; 4 5 6; 7 8 9]```
```>>> B```	```B =```
```array([[1, 2, 3],```	``` 1 2 3```
``` [4, 5, 6],```	``` 4 5 6```
``` [7, 8, 9]])```	``` 7 8 9```
```>>> numpy.mean(B)```	```>> mean(B)```
```5.0```	```ans =```
```>>> numpy.mean(B,0)```	``` 4 5 6```
```array([4., 5., 6.])```	```>> mean(B,2)```
```>>> numpy.mean(B,1)```	```ans =```
```array([2., 5., 8.])```	``` 2```
	``` 5```
	``` 8```

Note that every time you assign something to memory that was already previously named, like here, you overwrite the previous contents. We had already assigned the matrix "B" before, but now we reassigned it with different contents (contents that make it more clear what is going on). Note also that the default way to take the mean is column-wise, across rows. To get the row-means (across columns), we have to specify this by giving a second input to the mean function, the "2" denoting that we want to take the mean across the 2nd dimension—columns in MATLAB ("1" for Python)—this generalizes to higher-dimensional arrays.

Similarly, linearization works column-wise MATLAB takes the first column and stacks it on top of the second, which is in turn stacked on top of the third. If you want to linearize row-wise, you have to transpose the source matrix first—and assign the result to a temporary variable, here named "C," then linearize *that*. In Python, you can chain together the functions so that B is transposed and then flattened:

Python	MATLAB

`>>> B.T.flatten() array([1, 4, 7, 2, 5, 8, 3, 6, 9])`	`>> B(:)'`
	`ans =`
	`     1     4     7     2     5     8`
	`3     6     9`
	`>> C = B'; C(:)'`
	`ans =`
	`     1     2     3     4     5     6`
	`7     8     9`

Note that in the interest of space, we have transposed the outputs again, so that they yield a row vector; we don't want to fill half a page with this. But it isn't necessary for the logic that we try to illustrate here if you want to try it yourself.

WHAT CAN WE DO WITH ALL OF THIS?

To illustrate the power of representing data in matrix format, we will use several matrix operations. Matrix operations are particularly useful to solve problems in linear algebra. For instance, if you want to represent the average firing rate of a given kind of neuron with a matrix and the number of neurons you have of that particular kind in another matrix, you could calculate the total number of spikes fired (the expected spike count per second) in a given system very efficiently. Here, we model the nervous system of *Caenorhabditis elegans*, which has 302 neurons divided into seven different classes.

(For the sake of exercise, we pretend that *C. elegans* neurons produced action potentials and that they are easily classifiable in this way, which is not the case (Lockery & Goodman, 2009). But the number of neurons is not in dispute.).

Python	MATLAB
```>>> neuralFiringRates = numpy.array([0.5, 1, 2, 4, 8, 0.25, 3])``` ```>>> numberOfNeurons = numpy.array([45, 15, 35, 20, 50, 36, 101])``` ```>>> expectedSpikeCount = numpy.inner(neuralFiringRates,numberOfNeurons)``` ```>>> expectedSpikeCount``` ```899.5```	```>>neuralFiringRates = [0.5 1 2 4 8 0.25 3];``` ```>>numberOfNeurons = [45; 15; 35; 20; 50; 36; 101];``` ```>>expectedSpikeCount = neuralFiringRates*numberOfNeurons``` ```expectedSpikeCount =``` ``` 899.5```

The dot- or inner product corresponds to an element-wise multiplication of the two matrices, then a summing of the results.

Matrix operations are so native to MATLAB that you need to specify if you *just* want element-wise multiplication (as most people who are not doing linear algebra will want). Thus, avoiding doing a dot product is (somewhat ironically) achieved by putting dot operator immediately before the multiplication sign.

Python	MATLAB
```>>> expectedSpikeCount = neuralFiringRates*numberOfNeurons``` ```>>> expectedSpikeCount``` ```array([22.5, 15. , 70. , 80. , 400. , 9. , 303.])```	```>> expectedSpikeCount = neuralFiringRates.*numberOfNeurons'``` ```expectedSpikeCount =``` ``` Columns 1 through 3``` ``` 22.5 15 70``` ``` Columns 4 through 6``` ``` 80 400 9``` ``` Column 7``` ``` 303```

This yields the expected spike count per class of neuron, without adding them all up. Note that we had to transpose the MATLAB vector representing the number of neurons per class, as for element-wise computations, the number of columns and rows has to match.

The matrix multiplication is not commutative, so A*B is not B*A. One represents the inner (or dot) product, the other the outer product (helpfully referred to as the "tensor product") in mathematics:

Python	MATLAB
>>> spikeCountCrossComparison= numpy.out er(neuralFiringRates,numberOfNeurons)	>> spikeCountCrossComparison = numberOfNe urons*neuralFiringRates;

The output of this computation represents a 7 × 7 table that contains the full cross (all possible multiplications) of all possible firing rates times all possible neuron groups. This indicates what kinds of spike counts you could expect for a given neuron class if they fired at the rate that the other neuron groups do, which gives you a sense of where an empirical spike count from the entire *C. elegans* nervous system could fall (if *C. elegans* neurons fired action potentials and were subdivided into these groups). We suppressed the output here in the interest of space.

This approach would scale for a primate brain with billions of neurons. So you see the appeal of matrix operations with a "simple" system (for neuroscience, as simple as it gets).

It is not immediately obvious why you would want to do the outer product here, so let's give a more meaningful example. Say you want to study the visual system of *C. elegans* (of course *C. elegans* neither has eyes, nor a visual system, but the trend is definitely going toward simpler and simpler model organisms in visual neuroscience, on account of the genetic methods one can use—in conjunction with imaging methods to record data) and you want to check on the experimental conditions you create. You want to use five exposure durations and five light levels, fully crossed. The outer product of exposure durations and light levels yields the total photon count that the organism is exposed to, per condition:

Python	MATLAB

>>> exposureDurations = numpy.array([1,2,3,4,5])
>>> lightLevels = numpy.array([1,2,4,8,16])
>>> photonCounts=
numpy.outer(exposureDurations,lightLevels)
>>> photonCounts
array([[1, 2, 4, 8, 16],
       [2, 4, 8, 16, 32],
       [3, 6, 12, 24, 48],
       [4, 8, 16, 32, 64],
       [5, 10, 20, 40, 80]])
	>> exposureDurations = [1; 2; 3; 4; 5];
	>> lightLevels = [1 2 4 8 16];
	>> photonCounts = exposureDurations*light
	Levels
	photonCounts =
	  1 2 4 8 16
	  2 4 8 16 32
	  3 6 12 24 48
	  4 8 16 32 64
	  5 10 20 40 80

THE FIND FUNCTION

The `find` function is extremely useful to find the indices of matrix elements that satisfy certain logical conditions in MATLAB. The analogous function in Python is numpy's *where* function. For instance, say you (as the experimenter) are only interested in neurons that fire at a rate of more than three spikes per second. It's easy to just look at the matrix representing this simple nervous system in *C. elegans*. Good luck doing that for a primate brain. That's where `find` comes in. We first find the neuron classes that meet the condition, then their corresponding numbers and add them up. This allows us to find the number of neurons that meet our conditions.

Python	MATLAB

```>>> indices = numpy.where(neuralFiringRates>3)```
```>>>indices```
```(array([0, 0]), array([3, 4]))```
```>>> neuronsMeetingConditions = numpy.sum(numberOfNeurons[indices])```
```>>> neuronsMeetingConditions```
```70```
	```>> Indices = find(neuralFiringRates>3)```
	```Indices =```
	``` 4 5```
	
	```>>neuronsMeetingConditions =```
	```sum(numberOfNeurons(Indices))```
	
	```neuronsMeetingConditions =```
	``` 70```

ADDING MATRICES AND DEALING WITH HOLES IN ARRAYS

Say you run a reaction time experiment and you have data (in milliseconds) from two participants (of an unspecified species). They did four trials each:

Python	MATLAB
```>>> rtP1 = numpy.array([1000, 1500, 500, 1200])``` ```>>> rtP2 = numpy.array([2000, 1700, 3000, 700])```	```>> rtP1 = [1000 1500 500 1200];``` ```>> rtP2 = [2000 1700 3000 700];```

You can declare these two vectors representing the reaction time of each participating organism individually. But you would probably want to combine them in a matrix representing the reaction times (per trial) of these (and all future) participants. Note that with Python's numpy, we use the function `vstack`, which results in the vertical stacking of the arrays (the alternative being `hstack`, which horizontally stacks the arrays).

Python	MATLAB			

```>>> RTs=numpy.vstack((rtP1,rtP2))```
```>>> RTs```
```array([[1000, 1500, 500, 1200],```
```[2000, 1700, 3000, 700]])```
	```>> RTs = [rtP1;rtP2]```
	```RTs =```			

	1000	1500	500	1200
	2000	1700	3000	700

In this matrix, rows represent participants, columns trial numbers, and the numbers themselves the reaction time for a given participant and trial in milliseconds. Such a representation is very helpful to efficiently compute, for instance, the effects in learning experiments, as you can now use a function, e.g., the mean function to compute the average reaction time per trial or (if you are interested in whether some participants are simply faster than others) per participant.

So first, we'll compute RT per trial (across rows), then per participant (across columns)

Python	MATLAB			

```>>> numpy.mean(RTs,0)```
```array([1500., 1600., 1750., 950.])```
```>>> numpy.mean(RTs,1)```
```array([1050., 1850.])```
	```>> mean(RTs,1)```			

	1500	1600	1750	950
	```>> mean(RTs,2)```			
	1050			
	1850			

That's great, but there is a catch. Concatenating matrices like this only works if the number of elements per row and column in the matrix matches. Put differently, POM abhors holes in matrices. Matrices cannot have holes. This is a challenge because as a neural data scientist, you will frequently be confronted with missing data.

For instance, the participants were supposed to do four trials each, but let's pretend that one participant didn't respond during one of the trials:

Python	MATLAB

```>>> rtP2 = numpy.array([2000,1700,3000])```
```>>> RTs= numpy.vstack((rtP1,rtP2))```
```--```
```Traceback (most recent call last)```
```<ipython-input-100-4a3c856b8924> in <module>()```
```----> 1 numpy.vstack((rtP1,rtP2))```
```ValueError: all the input array dimensions except for```
```the concatenation axis must match exactly```
	```>> rtP2 = [2000 1700 3000];```
	```>> RTs = [rtP1;rtP2]```
	```Error using vertcat```
	```Dimensions of matrices being concatenated```
	```are not consistent.```

POM throws an error because it does not know how to align the successive rows. The way to deal with this is perhaps somewhat exotic, but it is best dealt with right up front, as it is so common. Briefly put, we have to represent the hole with a placeholder. The symbol for this is "NaN." This looks like a string of characters, but it is internally represented as a number. Which number? No number, as "NaN" stands for "not a number." That's the paradox, as we represent a value that is not a number numerically. But it is very helpful.

So let's indicate that the second participant missed the fourth trial by denoting it as nan. POM will understand. In case you are curious, the function isnan can detect whether something is not a number, but represented as a number:

Python	MATLAB
```>>> rtP2 = numpy.array([2000,1700,3000,float('NaN')])```	```>> rtP2 = [2000 1700 3000 nan];```
```>>> numpy.isnan(rtP2)```	```>> isnan(rtP2)```
```array([False, False, False, True], dtype=bool)```	``` 0 0 0 1```
	```>> isnumeric(rtP2)```
	``` 1```

There is a price to pay for this convenient representation. Any numeric operation that involves a nan will render the result a nan, regardless of other inputs. Think of a single drop of picrotoxin that can spoil an entire barrel of otherwise good wine:

Python	MATLAB
```>>> RTs = numpy.vstack((rtP1,rtP2))```	```>> RTs = [rtP1;rtP2]```
```>>> RTs```	```RTs =```
```array([[1000., 1500., 500., 1200.],```	``` 1000 1500 500 1200```
``` [2000., 1700., 3000., nan]])```	``` 2000 1700 3000 NaN```
```>>> numpy.mean(RTs,0)```	```>> mean(RTs,1)```
```array([1500., 1600., 1750., nan])```	``` 1500 1600 1750 NaN```
```>>> numpy.mean(RTs,1)```	```>> mean(RTs,2)```
```array([1050., nan])```	``` 1050```
	``` NaN```

As you can see, if there is a nan involved in a computation over columns, that entire column becomes nan. Similarly, if it involves rows instead, that entire row becomes nan.

Of course there are functions that deal with this, for instance nanmean, but the best way to address missing data is up front—detecting and eliminating ill-formed data. We will cover this in more detail later, but for now, we could do some basic data wrangling (find all the columns of the matrix where there is a nan, then eliminate them). As for data wrangling: Get used to it. Data wrangling (or data munging) takes up most of the time of the (neural) data scientist.

Python	MATLAB			
`>>> numpy.nanmean(RTs,0)`	`>> nanmean(RTs,1)`			
`array([ 1500., 1600., 1750., 1200.])`	1500	1600	1750	1200
`>>> numpy.nanmean(RTs,1)`	`>> nanmean(RTs,2)`			
`array([ 1050. , 2233.33333333])`	1050			
`>>> r,c=numpy.where(numpy.isnan(RTs)==1)`	2233.33333333333			
`>>> RTs=numpy.delete(RTs,c,r)`	`>> [r,c] = find(isnan(RTs)==1);`			
`array([[ 1000., 1500., 500.],`	`>> RTs(:,c) = [ ];`			
`       [ 2000., 1700., 3000.]])`	`>> mean(RTs,1)`			
`>>> numpy.mean(RTs,0)`	1500	1600	1750	
`array([ 1500., 1600., 1750.])`	`>> mean(RTs,2)`			
`>>> numpy.mean(RTs,1)`	1000			
`array([ 1000., 2233.33333333])`	2233.33333333333			

Note that, when eliminating elements, an entire row or column has to go. Deleting a row or column is done by assigning the empty matrix [ ] in MATLAB, or by using the `delete` function in `numpy`, where c indicates the index and r indicates the axis along which we are deleting. We chose to eliminate all data from the fourth column, not all data from the second participant. But it was our choice, we could also have eliminated the entire participant—if we wanted to be conservative and eliminate all participants who didn't respond in a trial.

Now that we got the absolute basics done, let us put them to good use by making a graph. We don't have data yet (that will come in the next chapter: Wrangling Spike Trains), so let's graph some random numbers. To get them, we need to reengage with the POM random number generator, albeit in a different way. Before, we drew numbers from a uniform distribution, now we do so from a normal distribution.

WHAT IS A NORMAL DISTRIBUTION? HOW DO WE DRAW FROM ONE, HOW DO WE PLOT ONE WITH POM?

A variable of interest tends to be normally distributed if the sample size is sufficiently large and if the generative process underlying it involves many independent but small effects. For instance, IQ is known to be normally distributed—the idea is that there are a lot of underlying but independent factors that contribute a little bit to the outcome (many genetic and environmental factors). Say you are interested in the relationship between brain size and IQ. IQ is distributed normally with a mean of 100 and a standard deviation of 15. You are curious what that would look like if one was drawing a sample of 10, then a sample of 10,000 from the population,

Pseudocode

1. Draw 10 numbers from a normal distribution with mean of 100 and standard deviation 15.
2. Open a figure
3. Plot a histogram of the sample distribution with sample size 10 with 20 bins
4. Show the figure

The plot in Fig. 2.1 is from Python.

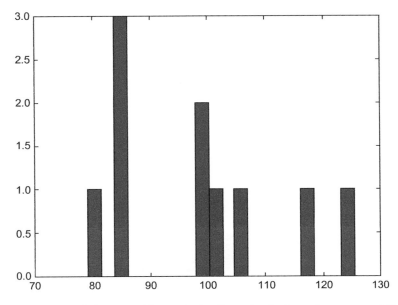

FIGURE 2.1   A histogram of IQ scores with sample size 10, randomly drawn from a normal distribution.

This figure is from Python, but the MATLAB figure will look very similar (albeit none of the figures will look exactly the same, even in Python because we plot random numbers).

Python	MATLAB
| ```
>>> sampleSize = 10
>>> meanIQ = 100
>>> stdIQ = 15
>>> sampleIQ = numpy.random.randn(sampleSize)*stdIQ+meanIQ
>>> import matplotlib.pyplot as plt
>>> f = plt.figure()
>>> ax = plt.hist(sampleIQ,bins=20)
>>> plt.show()
``` | ```
>> sampleSize = 10;
>> meanIQ = 100;
>> stdIQ = 15;
>> sampleIQ =
(randn(sampleSize,1).*stdIQ)+ meanIQ;
>> figure
>> histogram(sampleIQ,20)
>> shg
``` |

Not much of a normal distribution, is it? And so, the law of small numbers strikes again—it is a fact of life that small samples are extremely volatile, even if they are drawn from a known population distribution, like here. Sample distributions only start to resemble the population that they were drawn from, if the sample is sufficiently large. So let's try a sample size of 10,000—as we will see, sample size matters tremendously, even if one is drawing from the very same population, as we are.

Pseudocode

1. Draw 10,000 numbers from a normal distribution with mean of 100 and standard deviation 15
2. Open a figure
3. Plot a histogram of the sample distribution with sample size 10,000 with 20 bins
4. Show the figure

The plot in Fig. 2.2 is from Python.

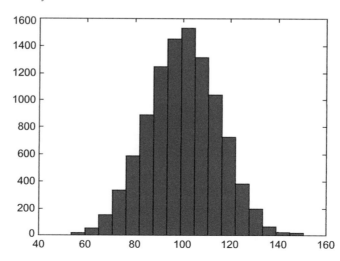

FIGURE 2.2   A histogram of IQ scores with sample size 10,000, randomly drawn from a normal distribution.

Python	MATLAB
```>>> sampleSize = 10000```	```>> sampleSize = 10000;```
```>>> sampleIQ = numpy.random.randn(sampleSize)*stdIQ+meanIQ```	```>> sampleIQ = (randn(sampleSize,1).*stdIQ)+meanIQ;```
```>>> f = plt.figure()```	```>>figure```
```>>> ax = plt.hist(sampleIQ,bins=20)```	```>> histogram(sampleIQ,20)```
```>>> plt.show()```	```>> shg```

Note that POM are *interpreted languages*. This means that they are executed (and interpreted) line by line at runtime. So once we redefine the sample size, we have to recalculate the distribution with the new sample size. The graph of the distribution with the new sample size is not updated automatically.

As you saw, even when drawing from a normal distribution, the result is unlikely to look like a "Gaussian" (or bell curve) when the sample size is small. This will become more important later when we talk about statistics in more detail.

HOW DO I PLOT SOMETHING MORE MEANINGFUL?

Remember how you created a matrix of total photon counts per imaginary condition for our experiment probing the visual system of *C. elegans*? Let's integrate much of what we learned in this chapter by assuming that we collected average firing rates in response to this visual stimulation (pretending for a moment that *C. elegans* can see, have a visual systems and neurons that generate action potentials). We can then vectorize the conditions and plot the imaginary firing rates of these neurons as a function of photon counts.

Pseudocode

1. Vectorize the photon counts
2. Determine which ones are unique to avoid plotting the same condition twice
3. Determine spike counts as function of photon counts, with random noise added—note that spike counts can only be integers, so we round
4. Open a figure
5. Plot the firing rates as a function of photon counts with linewidth 3
6. Label your axes

The plot in Fig. 2.3 is from MATLAB.

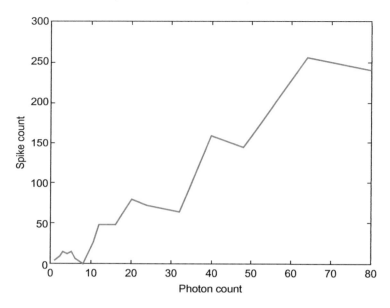

FIGURE 2.3   Spikes generated by the C. elegans nervous system as a function of photon count.

Python	MATLAB
```>>> photonCountsvector= numpy.unique(photonCounts.flatten()) >>> firingRates=photonCountsvector*numpy.round (5*numpy.random.rand(len(photonCountsvector)),1) >>> f = plt.figure() >>> ax = plt.plot(firingRates,lw=3) >>> plt.xlabel('photon count') >>> plt.ylabel('spike count') >>> plt.show()```	```>> photonCountsvector = unique(photonCounts(:)); >> firingRates = photonCountsvector.*round(5.*rand (length(photonCountsvector),1)); >> figure >> plot(photonCountsvector,firingRates,'linewidth',3) >> xlabel('photon count') >> ylabel('spike count')```

Of course it is more exciting to work with real neural data, not pretending what imaginary action potential counts would look like if *C. elegans* neurons generated action potentials and *C. elegans* could see. But we will do that in the Chapter 3, Wrangling Spike Trains. Even if we had neural data to give you (we do), you wouldn't know how to load into memory yet. Which brings us to...

HOW DO I SAVE WHAT I'M WORKING ON SO THAT I CAN LOAD IT AGAIN LATER?

Your variables are stored in RAM right now. This is called a "workspace" in MATLAB and an "environment" in Python. To see what they are named, how much memory they take up and other information (like class), type whos. If you lose power now or restart the machine, you will lose everything in RAM. To store the hard work you did, you need to save your variables to a more permanent storage medium, like a hard drive. To do so, you simply indicate the filename to the function *save* in POM. This stores the entire workspace/environment with the filename that you give it (the default extension is .mat, for MATLAB workspace), in the current directory. Then clear the workspace (delete everything from memory). If you type whos now, there will be nothing in memory. The function load looks for the filename in the current directory (unless you specify a path in the string you pass as an input to the load function.

Python	MATLAB

>>> %save myFirstVariables 1-50
>>> quit()
$ ipython -i myFirstVariables.py
>>> meanIQ
100
	>> save('MyFirstVariables')
	>> whos
	
	>> clear all
	>> whos
	
	>> load('MyFirstVariables')
	>> whos

One note about strings: in MATLAB, strings are denoted by single scare quotes ' ' and displayed in purple. In Python, to save the environment, we use the percent letter "%" as a *"magic operator."* Magic operators are a way to use special commands in the iPython terminal. Note that the magic operators used in this chapter only work with iPython (and Jupyter) and may cause errors if you try to embed them in scripts outside of the shell (more on this in Appendix C). Here, the %save command will save the entire environment to the variable myFirstVariables.py, and will save the lines of code specified. Note that in this example, we say to save the lines 1 through 50. To demonstrate how to reload this environment, we *quit* the iPython session with quit(), and reenter the interactive session of iPython with the environment variable myFirstVariables.py that we just saved. MATLAB saves all variables as a default, unless you specify a particular variable name to save in additional inputs to the save function.

So you could legitimately argue that this chapter ends with magic. And this is just 0.01.

⌘.   *Pensee on the incremental nature of learning*: Well, perhaps not precisely 0.01, but we used 0.01 for the smallest increment of proficiency that is measurable. Naturally, this will change over time. It might follow Weber's law (Fechner, 1860) and is likely idiosyncratic—its magnitude varying within and between people as well as depending on context and the application. Similarly, a femtometer is a small distance to a particle physicist, whereas an astrophysicist would probably consider 100,000 miles a small distance. We call the smallest meaningful measurable unit of learning a sparkulum. If you worked through this chapter, we think you just achieved one.

⌘⌘.   *Pensee on how to generate random numbers in Python and MATLAB*: Simply put, you don't. If you want to do this, your only option is to acquire a hardware random number generator. These are usually based on nuclear decay, diode noise, or another (as far as we know) inherently stochastic process. It is worth noting that all "random" numbers in Matlab or Python are generated by a pseudorandom process which is entirely and 100% deterministic. There is nothing random about it all. There are many specific algorithms to create them in either language (usually involving the multiplication of something with a large Mersenne Prime). To get an intuition for the process, think of the digits of pi which are (as far as we can tell) infinitely plentiful and distributed randomly. To generate a random sequence of numbers from this, all you need is a starting point. Once a starting point has been designated, the following sequence will be a sequence of random digits. However, it is an entirely determined sequence. Anyone else who starts from the same starting point will have the exact same sequence. In practice, for most people generating "random" numbers with POM, the starting point will be picked by the state of the system clock, which yields an unpredictable sequence in most cases. Careful when relying on Mersenne Primes. They are safe in principle, but not if everyone uses the exact same one. Then, the NSA can just crack that one and read everything, which seems to have happened in 2015. Apparently, at some point the NSA noted that 40–60% of encryption on the internet used the exact *same* prime. It looks like they then proceeded to build a $1 billion dedicated machine to crack just this one prime. And it looks like they did so successfully (Adrian et al., 2015). If you just want to do science, you don't have to worry about any of this. We simply want to point out that the stakes are high and that the numbers generated are emphatically not random, but indistinguishable from random to the naive observer (but not necessarily the NSA).

⌘⌘⌘.   *Pensee on the proper representation of equality*: The original idea behind the equal sign was to depict two parallel lines. What could be more equal than lines that are parallel? However, the implementation of this idea was somewhat muddled, which inverted the original intention. The arrangement and orientation of the lines matters as well. As it is now, the upper line in the equal sign seems to be literally oppressing the lower one from its position of unearned privilege. In other words, the current version of the equal sign is more aptly understood as a depiction of insufferable oppression. If one wants to advocate (and depict) true equality, a side-by-side arrangement of parallel vertical lines makes more sense, like this: | |.

## PART II

NEURAL DATA ANALYSIS

CHAPTER

3

Wrangling Spike Trains

Neurons are peculiarly social. Some neurons are veritable chatterboxes, some are quite quiet, and some are even trying to silence others. Some neurons prefer to talk only to close neighbors, whereas others send messages to comrades in far distant regions. In this sense, all neuroscience is necessarily social neuroscience. Of course neurons don't communicate with each other via spoken words. Rather, they use action potentials (also known as voltage *spikes*) as their universal means of long-distance communication. Every neuron sends spikes in its own idiosyncratic fashion and every neuron uses its dendritic arbor to receive signals from other neurons in turn. The interface points between neurons are known as *synapses*. A neuron may have many points (anywhere from one for some neurons in the retina to more than 100,000 for neurons in the cerebellum) of communication (synapses) with other neurons. While spikes are not exchanged directly—the signal crossing the synapse is chemical in nature in almost all synapses—it is the voltage spikes that drive the neural communication machinery. Specifically, spikes traveling down the axon of the *presynaptic* neuron trigger the chemical action in the synapse that enact further voltage changes—and perhaps more spikes in the *postsynaptic* neuron. We will consider details of this signal exchange in Chapter 6, Biophysical Modeling and Chapter 7, Regression.

For now, we will take for granted that neurons use spikes as their preferred medium of communication. It is fair to say that a major challenge faced by contemporary neuroscientists is to elucidate the meaning of these spikes. Put differently, we (the neuroscience community) are trying to crack the neural code. Our starting point in this pursuit is the signal itself—the spikes. Due to their nature as all-or-none events, we will represent the occurrence of spikes over time as numbers, specifically zeroes for "no spike" or ones for "spike." So consider the following list of numbers:

[0,0,0,0,0,0,0,0,0,1,0,1,0,1,0,0,0,1,0,0,0]

Neural Data Science.
DOI: http://dx.doi.org/10.1016/B978-0-12-804043-0.00003-9

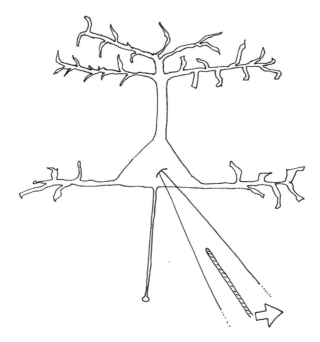

FIGURE 3.1   Extracellular electrode recording voltage spikes from an individual neuron in a dish.

By itself, this list of numbers is meaningless. However, let's assume that we have a neuron in a dish, that the neuron is alive, that it is capable of sending spikes, that this neuron will generally not send any spikes in the dark, and that this neuron will send some amount of spikes if you shine a green light (at a wavelength of 550 nm) on it. Let's also assume that we have a recording electrode near the point where this neuron sends spikes, and that our electrode has the fancy ability of telling our computer whether or not the neuron is spiking over time, as captured in the vector with 0's and 1's above. This scenario is schematized in Fig. 3.1.

Without knowing yet why the spikes occurred, we can make a couple of remarks about this list of numbers representing spikes. First, we know how many spikes are in the list:

Pseudocode	Sum up the numbers in the vector
Python	```>>> sum([0,0,0,0,0,0,0,0,0,1,0,1,0,1,0,0,0,1,0,0,0])``` ```4```
MATLAB	```>> sum([0,0,0,0,0,0,0,0,0,1,0,1,0,1,0,0,0,1,0,0,0])``` ```ans = 4```

But you didn't need POM to tell you that there are 4 ones in this list, as you can see that immediately. It is, however, convenient for us to measure how long the list is using the function *len* in Python and *length* in MATLAB:

Pseudocode	Count the number of elements in the longest dimension of the vector
Python	```>>> len([0,0,0,0,0,0,0,0,0,1,0,1,0,1,0,0,0,1,0,0,0])``` ```21```
MATLAB	```>> length([0,0,0,0,0,0,0,0,0,1,0,1,0,1,0,0,0,1,0,0,0])``` ```ans = 21```

Which means the list of numbers has 21 entries or elements. Recall that each number in the list represents whether or not the neuron sent a spike at that time, that the 0 on the far left of the list represents *time* = 0, and that each successive number on the list represents whether or not the neuron sends a spike at that time. We could say:

Pseudocode	Create a list of 21 successive integers representing time and align it with the 21 neuron states

Python	```range(21), [0,0,0,0,0,0,0,0,0,1,0,1,0,1,0,0,0,1,0,0,0]```
	```([0,1,2,3,4,5,6,7,8,9,10,11,12,13,14,15,16,17,18,19,20],```
	```[0,0,0,0,0,0,0,0,0,1, 0, 1, 0, 1, 0, 0, 0, 1, 0, 0, 0])```

MATLAB	```>> [linspace(0,20,21); [0,0,0,0,0,0,0,0,0,1,0,1,0,1,0,0,0,1,0,0,0]]``` ```ans =```

0	1	2	3	4	5	6	7	8	9	10
0	0	0	0	0	0	0	0	0	1	0
11	12	13	14	15	16	17	18	19	20	
1	0	1	0	0	0	1	0	0	0	

We interpret this to mean that at times 0, 1, 2, 3, 4, 5, 6, 7, and 8, the neuron is not spiking. At time 9, it spikes, at time 10 it is quiet, at time 11 it spikes, at time 12 it is quiet, at time 13 it spikes, at times 14 through 16 is it quiet, and then at time 17 it spikes one last time before being quiet again.

We said earlier that this neuron tends to spike if it is illuminated with green light, but not in darkness. What we are simulating here is a tool known as *optogenetics*, where neurons will actually increase their activity in response to light (Boyden et al., 2005; Tye & Deisseroth, 2012).

So let's indicate the time points during which such a green light was on in ***italic bold***, leaving the rest of the time points unbold (representing times during which the light was off):

[0,1,2,3,4,***5,6,7,8,9,10,11,12,13***,14,15,16,17,18,19,20]

Let us assume now that each time point above is in units of milliseconds. What this means is that we started recording from the neuron at time 0 when the light was off. After 4 ms of recording, on the fifth millisecond, the green light was turned on. The light then stays on for 9 ms (through the 13th millisecond) before shutting off.

With this knowledge of the stimulus conditions, we can determine a characteristic feature of this neuron: *"first spike latency to stimulus."* This parameter is generally used by neuroscientists to establish how "fast" or "ready to spike" any given neuron is.

We now know just enough about the stimulus and the neuron to be good neuroscientists and form a hypothesis. Let's hypothesize that the neuron always fires a spike 4 ms after a light is turned on.

Let's put the string of 0's and 1's into a *variable* now:

Pseudocode	Assign data to variable spikeTrain
Python	`>>> spikeTrain = [0,0,0,0,0,0,0,0,0,1,0,1,0,1,0,0,0,1,0,0,0]`
MATLAB	`>> spikeTrain = [0,0,0,0,0,0,0,0,0,1,0,1,0,1,0,0,0,1,0,0,0];`

With the spikes now stored in spikeTrain, we can pull out spikes at different time points. That is, we can ask, say, at time $t = 5$, when the light is first turned on, is the neuron spiking?

Pseudocode	Output the contents of variable spikeTrain at the position corresponding to $t = 5$
Python	`>>> spikeTrain[5]` `0`
MATLAB	`>> spikeTrain(6)` `ans = 0`

A resounding no. But what about if we want to know if the cell spikes at any time *after* $t = 5$?

Pseudocode	Output all elements of spikeTrain after the position corresponding to $t = 5$
Python	```>>> spikeTrain[5:]``` `[0,0,0,0,1,0,1,0,1,0,0,0,1,0,0,0]`
MATLAB	```>> spikeTrain(6:end)``` `ans = 0 0 0 0 1 0 1 0 1 0 0 0 1 0 0 0`

Note: Here we see some subtle but critical differences in Python versus MATLAB. First, in MATLAB, the first element of a vector is element "1," whereas the corresponding first element in Python is "0." So to access the same element (here the one at time 5), we have to add 1 to the MATLAB index. Second, the colon operator: returns all elements from a starting point until an endpoint. Python assumes you want all elements until the end of the vector if no endpoint is specified, the corresponding MATLAB command to specify "all elements until the end of the vector" is "end." Finally, note that the MATLAB command uses parentheses whereas the Python command uses square brackets.

This output (in POM) represents the neuron's spiking activity after the green light turned on. If the first output value in the list were a 1, then the neuron's *latency to first spike* would be 0 ms, i.e., the neuron's response would be coincident with the light turning on. But things rarely happen instantly in biology, let alone neuroscience. Rather, whatever makes our neuron spike in response to light takes some time to flip some internal switches before the spike occurs (we'll address the internal spiking mechanisms in chapter: Biophysical Modeling). For now, we just acknowledge that the neuron takes some time before spiking. The record of the successive spiking of a neuron over time is called the "*spike train*" (see Fig. 3.2) and various measures to characterize the spike train have been proposed.

The time it takes for a neuron to be responsive, i.e., for a neuron to spike in response to a stimulus (in our case, to the light) is known as the *response latency*. Different ways exist to measure response latency, but for a single light pulse and a single recording from a neuron that spikes relatively infrequently (or sparsely), the time it takes for the first spike to occur is a reasonable measure. So let's do that, and calculate the *latency to first spike* by typing:

Pseudocode	Find the first value in the elements of spikeTrain that matches 1
Python	```>>> spikeTrain[5:].index(1)``` `4`
MATLAB	```>> find(spikeTrain(6:end)==1); ans(1)-1``` `ans = 4`

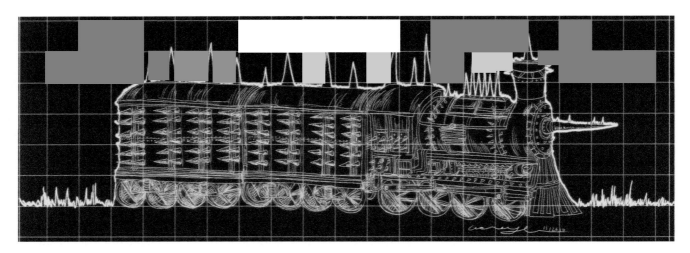

FIGURE 3.2 **A** *spike train*, as seen on an oscilloscope. Voltage spikes recorded from an individual neuron are plotted over time.

Note that in the MATLAB version of this code, we concatenate two commands in the same line by the use of the semi-colon operator. We have to bring the result into the same timebase as the Python code by adding 1 to the start point, then subtracting it again from the result. If we had started with calling the first element time "1," we would have had to subtract one from Python to get the right index, then add 1 to the answer. Whether the first element of a vector is element zero or element one is a philosophical consideration that we will further explore in Appendix A.

Here we took our list of values in the variable spikeTrain that occurred after a time of 5 ms and used the Python function *index* (*find* in MATLAB) to find the first value in this variable that represents the spike train that matches 1 (representing a spike). We can make this more flexible by giving the light onset time 5 a variable name, and also the spike indicator value 1 a variable name. Generally speaking, it is a bad idea to hard-code values that are given to functions explicitly. It is almost always better to use variables—code that uses variables is much easier to maintain.

Pseudocode	Assign values to variables, then pass those to the function to avoid hard-coding

| Python | ```
>>> lightOnsetTime = 5
>>> spikeValue = 1
>>> spikeTrain[lightOnsetTime:].index(spikeValue)
4
``` |
|---|---|
| MATLAB | ```
>> lightOnsetTime = 5;
>> spikeValue = 1;
>> mShift = 1;
>> find(spikeTrain(lightOnsetTime+mShift:end)==spikeValue);ans(1)-mShift
ans = 4
``` |

This version (without the hardcoding) also makes it more clear what is going on in the MATLAB case. Because of MATLAB's indexing conventions, we know that MATLAB indices are always shifted by 1 relative to Python. So we allocate this to a constant ("mShift") and add it to the lightOnsetTime, then subtract it again from the result. Note that the output is the same, which is reassuring, as our results should depend on the properties of the neuron we study, not the software we use to analyze the data we record from it.

Again, in order to find the latency to first spike, we take the variable *spikeTrain*, which contains all of the 1's and 0's that represent the presence or absence of spiking at a given time bin and look only at the times after the light onset, then look for the first time that the vector after light onset contains a 1. Technically speaking, this command returns the number of bins between light onset and the first spike, but as we know that the bin width is 1 ms, we can interpret the result as a time—the latency is 4 ms. In order to be able to reuse a result that we just calculated in this fashion, we should assign the results of a command to a variable. In this case, we will call it "latencyToFirstSpike." Generally speaking, it is advisable to use variable names that are meaningful as that makes code much more readable.

Pseudocode	Calculate latency to first spike and assign it to a variable with meaningful name
| Python | ```
>>> latencyToFirstSpike = spikeTrain[lightOnsetTime:].index(spikeValue)
>>> print latencyToFirstSpike
4
``` |
| MATLAB | ```
>> temp = find(spikeTrain(lightOnsetTime+mShift:end)==spikeValue);
>> latencyToFirstSpike = temp(1)-mShift
latencyToFirstSpike = 4
``` |

Note that in the example above, the MATLAB computations are done in two steps. We first declare a temporary variable temp that finds *all* instances in which spikes occur after light onset. We then, in a second step, find the index of the first element and correct for the MATLAB indexing shift by subtracting one, and assigning that to the variable latencyToFirstSpike.

Python has its own set of idiosyncrasies—these brief calculations in MATLAB involve only regular parentheses. In Python, *square brackets*, [ ], are used when specifying the range of values within a list of numbers. In contrast, regular parentheses, ( ), are used to invoke functions with the particular parameters within the parentheses as inputs. In this case, we pass the variable spikeValue to the function *index*, which is a built-in function—*index* is not the only such function. Python has many functions that we'll be using, and when we do, we'll use parentheses to give them values to operate on.

"I should make a plot"

FIGURE 3.3   Cartoons can efficiently illustrate important principles.

Now we have our estimate of the neuron's latency (4 ms)! As the readers of your research papers are likely to be primates and primates are predominantly visually guided animals, we should make a plot to illustrate the spiking activity of the neuron (Fig. 3.3).

To plot the spiking activity, we need to know the time of every spike in the list. People sometimes called these the *spike timestamps* but we'll just call them *spikeTimes*:

Pseudocode	Find and then output the times at which the neuron spikes

Python	```>>> spikeTimes = [i for i,x in enumerate(spikeTrain) if x==1]```
	```>>> print spikeTimes```
	```[9, 11, 13, 17]```

MATLAB	```>> spikeTimes = find(spikeTrain==1)-mShift``` ```spikeTimes = 9 11 13 17```

The Python part of this is a whopping nest of code! Let's untangle it a bit. First, see how we put the list of numbers `spikeTrain` (a bunch of 1's and 0's) into the function `enumerate`. Don't bother typing `enumerate(spikeTrain)` into your command line or trying to print it yet. The function `enumerate(spikeTrain)` cycles through the list `spikeTrain` and keeps track of the index of each element in `spikeTrain`.

The middle part of the code `i,x in enumerate(spikeTrain)` means that we will be going through each element in `spikeTrain` and naming each element along the way "x," and wherever "x" is in the list `spikeTrain`, we'll call that location "i."

A diagram might help:

i will successively be each element in: [0,1,2,3,4,5,6,7,8,9,10,11,12,13,14,15,16,17,18,19,20]
x will successively be each element in: [0,0,0,0,0,0,0,0,0,1, 0, 1, 0, 1, 0, 0, 0, 1, 0, 0, 0]

Let's look at that line of code again:

```
>>> spikeTimes = [i for i,x in enumerate(spikeTrain) if x==1]
```

Note that POM make use of double equals signs to test for the equality of two values. We have already encountered this before, in Chapter 2, From 0 to 0.01, but it is worth repeating (in our experience, this is confused a lot, particularly in the beginning) that a single equal sign is an assignment operator, assigning whatever is on the right (usually the result of a computation) to the left (usually a variable).

We now understand that the line means to give us the indices of *i* where *x* is equal to 1.

i is [0,1,2,3,4,5,6,7,8,9,10,11,12,13,14,15,16,17,18,19,20]
and
x is [0,0,0,0,0,0,0,0,0,1, 0, 1, 0, 1, 0, 0, 0, 1, 0, 0, 0]

As you can see, this is the case for the indices *i* of 9, 11, 13, and 17. And now, given this line, Python can see it too and return it to you (and put it into the variable "spikeTimes" as well).

We can conceive of this problem in reverse, too.
If *x* is equal to 1, *where x is each element in* [0,0,0,0,0,0,0,0,0,1, 0, 1, 0, 1, 0, 0, 0, 1, 0, 0, 0]
give us the corresponding value in *i*:

[0,1,2,3,4,5,6,7,8,9,10,11,12,13,14,15,16,17,18,19,20]

We underlined the *i*'s and *x*'s where *x* is equal to 1.

That's a lot of work (the equivalent MATLAB code seems much simpler) and would be simpler yet if we didn't have to bring the result into a format that conforms to Python's zero-indexing conventions—but hey, that's a small price to pay to use a real programming language, like the real hackers do. After all, social life is all about tribal signaling.

But let's not lose sight of the fact that the point of doing this was so that we can graph it. To graph (or "plot," as we will call it from now on) something in Python, we need to import a library of functions designed for this purpose. This library is called "*matplotlib*," a library that gives us all the plotting tools we need at this point.

Pseudocode	Import library of plotting functions
Python	>>> import matplotlib.pyplot as plt
MATLAB	No equivalent. These plotting functions already come with MATLAB

This means we've loaded the function *pyplot*, a function of the library *matplotlib*, but when we loaded it we gave it a shorthand (*plt*) so we can refer to it more easily—and with a nickname of your choice! You'll find that the strategy of importing functions as two or three letter shorthands will save you a lot of typing. We recommend it.

We will now create a figure to work on:

Pseudocode	Open figure
Python	>>> fig = plt.figure()
MATLAB	>> figure

This establishes a frame to work within. Within this frame, we can have one to many smaller windows, or *subplots*. For now, we're plotting a small set of spikes, so we only need one *subplot*.

Pseudocode	Place a subplot (or axes) into the figure
Python	>>> ax=plt.subplot(111)
MATLAB	No equivalent. If there is only one pair of axes, it will take up the entire figure by default. You could still type subplot(1,1,1) and place axes, but this is not necessary.

What the *111* means will be a little more clear once we have multiple subplots in a figure. For now, just know that if you want only one plot in your figure, you are saying that you want one large subplot, and therefore use *subplot(111)*. We call this subplot "ax" in honor of the axes of a Cartesian plot, which is the kind we want here.

To plot the spike times, we'll use the common visualization method of a *spike raster plot*. In such a plot, each spike is represented as a vertical line (at the time when it occurred, with time on the *x*-axis) (Fig. 3.4).

Pseudocode	Plot vertical lines at the times when a spike occurred, then show the figure
Python	>>> plt.vlines(spikeTimes, 0, 1) >>> plt.show(block=False)
MATLAB	>> line(repmat(spikeTimes,2,1),repmat([0; 1],1,4),'color','k') shg

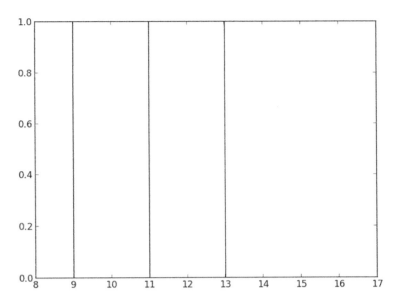

FIGURE 3.4   A bare bones raster plot of a single trial.

Voila, our first plot of neuroscience data. Obscenely, we have neither axis labels nor units … yet. Before we rectify this problem, let's discuss how the figure was brought about by the commands directly above.

In Python, we invoke the function *vlines* (a subfunction of the plotting package we called *plt*) by putting a period between them. This is generally the way to invoke subfunctions of imported packages. We then use parentheses to give values to this function. The three arguments we pass to the function *vlines* are: First a list of spike times, then the minimum value of the vertical lines—0, then the maximum value of the vertical lines—1.

In MATLAB, there is no function for vertical lines specifically, so we use the general purpose function *line*. It takes matrices to specify *x* and *y* coordinates and plots one line for each column whereas row-values indicate start and end positions of the lines. We have to create these matrices first, so we create two 2 × 4 matrices with the *repmat* function.

We also have to set the color of the lines to black as the default MATLAB color is blue in order to match the plot produced by Python. The function *shg* shows the current graph.

This plot illustrates when the neuron spikes, but doesn't contain any information about the light stimulus yet. Let's make the time during which the light is on a shaded green:

Pseudocode	Add a shaded green rectangle from times 5 to 14
Python	In :plt.axvspan(5,14,alpha=0.1,color='g')
MATLAB	rectangle('Position',[5,0,9,1],'FaceColor',[0.7 1 0.7],'linestyle','none')

This creates a shaded green box that spans the figure vertically and is bounded horizontally at 5 and 14.
Python: The *alpha* value makes the box transparent (smaller *alpha* values make the color appear lighter).
MATLAB: Specifies the color of the box by giving it an RGB triplet. A light green in this case.
Let's now specify the range of times our *x*-axis should span by invoking the *xlim* function so that we can see times during which the neuron did not fire, i.e., before visual stimulation:

Pseudocode	Setting the range of the x-axis to include the entire time interval of interest
Python	In :plt.xlim([0,20])
MATLAB	xlim([0 20])

Before showing this figure to anyone, we strongly recommend to add a label to the *x*-axis and a title to the figure.

Pseudocode	Add meaningful axis and figure labels, which is critical in science
Python	>>> plt.title('this neuron spikes in response to a single light stimulus') >>> plt.xlabel('Time (in milliseconds)')
MATLAB	>> title('this neuron spikes in response to a single light stimulus') >> xlabel('Time (in milliseconds)')

Due to the noise inherent in the system as well as the measurements, data from a single trial are rarely sufficient to reach reliable conclusions about the connection between visual stimulation and the neural response (Fig. 3.5).

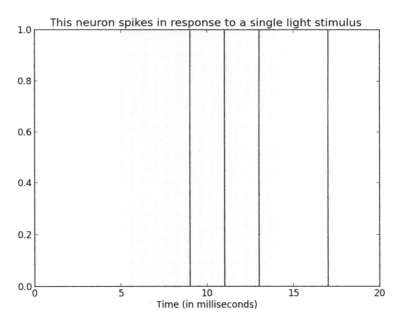

FIGURE 3.5   A raster plot of a single trial with axis labels, title, and stimulation condition.

What if there is some mechanism inside the neuron that causes it to spike highly unreliably? What if we are recording the signals of this neuron in a noisy environment? In theory, we would need to stimulate the neuron an infinite number of times and record an infinite number of responses in order to be really sure. But most people are not theorists, living in a nonplatonic world. As such (and neuroscientists at that) we have to make do with less than infinite amounts of data. Say we had just enough funding to allow us to record data from 10 spike trains and 10 identical (green) light stimuli. These data are contained in the "tenSpikeTrains" variable. It can be downloaded from the companion website, but we also print it on the next page.

Pseudocode	Representing the ten spiketrains in the respective formats

Python	`>>> tenSpikeTrains = [[0,0,0,0,0,0,0,0,0,1,0,1,0,1,0,0,0,1,0,0,0],[0,0,0,0,0,0,0,0,1,1,0,0,0,1,0,0,`
	`0,1,0,0,0],[0,1,0,0,0,0,0,0,0,1,0,0,1,0,0,0,0,0,1,0],[0,0,0,0,0,0,0,0,1,1,0,1,0,0,0,0,0,0,1,0,0],`
	`[0,0,0,0,0,0,0,0,1,0,0,1,0,0,0,0,0,0,0,0],[0,0,0,0,0,0,0,0,0,1,1,0,0,0,1,0,0,0,1,0],[0,0,0,0,0,`
	`0,0,0,1,1,1,0,0,1,1,0,0,1,1,0,0],[1,0,0,0,0,0,0,0,0,0,1,1,0,0,0,0,1,0,0,0,0],[0,0,0,0,0,0,0,0,1,1,0`
	`,0,1,1,0,0,1,1,0,0,0],[0,0,0,0,0,0,1,0,0,1,0,1,0,0,0,0,0,1,1,0,0]]`

MATLAB analogous	`>> tenSpikeTrains = {[0,0,0,0,0,0,0,0,0,1,0,1,0,1,0,0,0,1,0,0,0],[0,0,0,0,0,0,0,0,1,1,0,0,0,1,0,0,0`
	`,1,0,0,0],[0,1,0,0,0,0,0,0,0,1,0,0,1,0,0,0,0,0,1,0],[`
	`0,0,0,0,0,0,0,0,1,0,0,1,0,0,0,0,0,0,0,0],[0,0,0,0,0,0,0,0,0,1,1,0,0,0,1,0,0,0,1,0],[0,0,0,0,0,0`
	`,0,0,1,1,1,0,0,1,1,0,0,1,1,0,0],[1,0,0,0,0,0,0,0,0,0,1,1,0,0,0,0,1,0,0,0,0],[0,0,0,0,0,0,0,0,1,1,0,`
	`0,1,1,0,0,1,1,0,0,0],[0,0,0,0,0,0,1,0,0,1,0,1,0,0,0,0,0,1,1,0,0]}`

MATLAB suitable	`>> tenSpikeTrains = [0,0,0,0,0,0,0,0,0,1,0,1,0,1,0,0,0,1,0,0,0; 0,0,0,0,0,0,0,0,1,1,0,0,0,1,0,0,0,1`
	`,0,0,0;0,1,0,0,0,0,0,0,0,1,0,0,1,0,0,0,0,0,1,0;0,0,0,0,0,0,0,0,1,1,0,1,0,0,0,0,0,0,1,0,0;0,0,0,0,0,`
	`0,0,0,0,1,0,0,1,0,0,0,0,0,0,0;0,0,0,0,0,0,0,0,0,1,1,0,0,0,1,0,0,0,1,0;0,0,0,0,0,0,0,0,1,1,1,0`
	`,0,1,1,0,0,1,1,0,0;1,0,0,0,0,0,0,0,0,0,1,1,0,0,0,0,1,0,0,0,0;0,0,0,0,0,0,0,0,1,1,0,0,1,1,0,0,1,1,0,`
	`0,0;0,0,0,0,0,0,1,0,0,1,0,1,0,0,0,0,0,1,1,0,0]`

Note that Python uses nested square brackets here. It appears, from the placement of the *square brackets*, that there are 10 lists (each inside *square brackets*) nested inside of one big, all-encompassing list (note the double *square brackets* at the very beginning and end of *tenSpikeTrains*). In fact, this is the case and *tenSpikeTrains* is technically a list of lists. We can represent these spiketrains like that in MATLAB too, by using cells (the "MATLAB analogous" code), but the most suitable way to represent these data in MATLAB is as a 10 × 21 matrix ("MATLAB suitable"). In other words, we represent each spiketrain as entries in 21 consecutive columns and each individual spiketrain as a separate row. So at every point in the MATLAB suitable code above where there is a],[in Python, there is a semicolon in MATLAB. This works because the time base (21 bins with a width of 1 ms) is the same for each spiketrain. If this was not case, e.g., if there were missing data, using cells would be more apt. Matrices are ordered arrangements of numbers and cells are ordered arrangements of matrices. So Cell:Matrix as Matrix:Number, in MATLAB. Cells can accommodate matrices with different dimensionalities in each entry so they are very helpful, but make it harder to do some computations on them. In addition, they complicate the notation—note the curly braces { } that indicate we are dealing with cells. We will not say more about them at this point, but will when we have to, later in this book.

Back to the data, if we look at the first entry of *tenSpikeTrains*, we find:

Pseudocode	Return the contents of the first spike train
Python	```>>> tenSpikeTrains[0]``` ```[0,0,0,0,0,0,0,0,0,1,0,1,0,1,0,0,0,1,0,0,0]```
MATLAB analogous	```>> tenSpikeTrains{1}``` ```ans = 0 0 0 0 0 0 0 0 0 1 0 1 0 1 0 0 0 1 0 0 0```
MATLAB suitable	```>> tenSpikeTrains(1,:)``` ```ans = 0 0 0 0 0 0 0 0 0 1 0 1 0 1 0 0 0 1 0 0 0```

The first spiketrain is contained in the first row (MATLAB suitable) or first cell (MATLAB analogous) or the 0th list (Python). This is the flipside of Python zero indexing. It made sense that the first time of a time series would be time zero. But the first element of a list is still the first element, not the zeroth element. So each indexing convention (0 for Python, 1 for MATLAB) has use cases where it is more "natural," which is why both of them are still around.

Regardless of implementation, the command doesn't return a single value, but a list. How many elements are in this master list?

Pseudocode	How many elements does the variable tenSpikeTrains have?
Python	```>>> len(tenSpikeTrains)``` ```10```
MATLAB analogous	```>> length(tenSpikeTrains)``` ```ans = 10```
MATLAB suitable	```>> size(tenSpikeTrains,1)``` ```ans = 10```

In the "MATLAB suitable" case, we have to count the number of rows, which is achieved by telling the *size* function to evaluate the first dimension (rows).

You might have expected way more than that (like around 200), but the *len* function in Python looks at how many things are inside the list much in the same way that a bundle of bananas only counts as one item at the grocery store when you are in the express lane. It simply returns the number of elements (in this case lists) it contains. The same is true for the MATLAB analogous case—the cell has 10 entries, each of which is a matrix (and each of which represents an individual spike train). Accessing cell contents and doing computations on them is more involved than is worth getting into at this point, so we'll punt that to Chapter 4, Correlating Spike Trains, and continue solely with the "suitable" case below.

In order to make a raster plot for the data from *all* the trials, we use a similar approach we used for plotting the data from a single trial, except we cycle through each list in the list. To do this, we will use a *for loop*.[a]

Python	MATLAB (suitable)
| ```
>>> fig = plt.figure()
>>> ax = plt.subplot(1,1,1)
>>> for trial in range(len(tenSpikeTrains)):
...     spikeTimes = [i for i,x in enumerate
(tenSpikeTrains[trial]) if x==1]
...     plt.vlines(spikeTimes,trial,trial+1)
>>> plt.axvspan(5,14,alpha=0.1,color='g')
>>> plt.xlim([0,20])
>>> plt.show()[b]
``` | ```
>> ax=figure
>> rectangle('Position',[5,0,9,11],'FaceColor', ...
[0.7 1 0.7],'linestyle','none')
>> for ii = 1:size(tenSpikeTrains,1)
    >> spikeTimes = find(tenSpikeTrains(ii,:)==1)-mShift;
    >> line(repmat(spikeTimes,2,1),repmat([ii-0.5;ii+0.5] ,1, ...
length(spikeTimes)),'color','k')
>> end
>> xlim([0 20])
>> shg
``` |

As for Python, there are a few noteworthy things about this code, starting with the *for loop*. We know that the length of *tenSpikeTrains* is 10, so we can think of the *for loop* line of code as:

[*aside: We put this line of code front and center to emphasize it and talk about, not for you to retype it into the command line, or think that code usually sits centered in the page*]

```
for trial in range(10):
```

We can also see that range(10) is:

```
>>> range(10)
[0,1,2,3,4,5,6,7,8,9]
```

---

[a] *For loops* are your friend for easy plotting, and your enemy in heavy computation. If you are cycling through a network simulation or calculation, too many nested *for loops* will bring even the most powerful computer to its knees. As you begin to learn more programming tools, always be asking yourself if the code can be written without use of a *for loop*. But we are introducing the concept just now—we'll cover later how to write code that runs faster. Right now, using a for loop is fine, particularly as it is usually much clearer to understand what is going on when using loops.

[b] You may notice that if you use the command plt.show() that you will be unable to continue coding. To be able to continue coding, use the command: plt.show(block=False). To update the figure with any additional commands (for example, if after you show the figure you want to add an axis label) use plt.draw() after your additional command to update the figure.

That's right, range(10) is equal to the list [0,1,2,3,4,5,6,7,8,9]. In MATLAB, the command 1:size(tenSpikeTrains,1) achieves exactly the same.

We can thus think of the *for loop* line of code in Python:

```
for trial in range(len(tenSpikeTrains)):
```
as equivalent to
```
for trial in [0,1,2,3,4,5,6,7,8,9]:
```

Which means we are going to go to the next line of code after the *for loop* 10 times, and each time we go to the next line the variable *trial* will increase to the next value in the list (starting with 0). It is called a *for loop* because it loops through a list of values. The principles of the for loop in Python are reflected in the for loop used in MATLAB:

```
for ii = 1:size(tenSpikeTrains,1)
```

So whats up the line after the *for loop*? There's the ellipses and some spaces before we get to the *spikeTimes* line, which is very similar to how we got *spikeTimes* before. Note in Chapter 2, From 0 to 0.01, we discussed the importance of *white space* in Python. If you are running iPython from a command line or terminal, it will automatically create an indented next line if the preceding line ends with a colon (see examples of frequently made mistakes when using loops in the terminal in Appendix B). Ellipses are a placeholder for whitespace. In effect they represent white space explicitly.

```
>>> for trial in range(len(tenSpikeTrains)):
...     spikeTimes = [i for i,x in enumerate(tenSpikeTrains[trial]) if x==1]
...     plt.vlines(spikeTimes,trial,trial+1)
```

It will keep creating indented lines and staying inside the grasp of the *for loop* until you hit *return* on a blank line. This indicates to Python that the for loop has ended. Notice how the line of code

```
>>> spikeTimes = [i for i,x in enumerate(tenSpikeTrains[trial]) if x==1]
```

is similar to how we got `spikeTimes` for a single trial. This time though, we have our 1's and 0's within lists inside of the list `tenSpikeTrains`, so the variable `trial`, which iterates through the values *0–9*, will call each of the *10* lists inside the list `tenSpikeTrains`. `spikeTimes` is now a temporary variable, and each time `trial` is updated in the for loop, `spikeTimes` is reassigned to a new list of spike times.

Let's reiterate what `enumerate` does. This uniquely Python function allows us to build a for loop within a single line of code, as `enumerate` returns both each value in the list (like the `for` loop above) as well as the index of the value, where `i` is the index, and `x` is the value. Additionally, within this single line we condition returning the index `i` on the value `x` equaling 1. Thus, in this single line of code, we return the indices of the values equal to one, in a Pythonic fashion known as a *list comprehension*.

The MATLAB code tries to achieve the same, but note that the ellipses (…) are at the end of the line. They indicate that the command continues in the next line. So the use of ellipses is very different in Python versus MATLAB. In Python, indicating that the command continues in the next line is done via a backslash, "\."

For each trial in tenSpikeTrains we plot vertical lines for the spikes, where the values *trial, trial+ 1,...,trial+ 9* are stacked sequentially and vertically, where every "row" of the plot represents a trial.

Both POM have commenting conventions, which we are starting to use here—and from now on. Anything after the hashtag (#) in Python isn't read as programming code and is ignored by the computer, meaning it is not interpreted as an instruction and thus not attempting to execute it. In general, we recommend to write a sentence at the beginning of a paragraph of code to explain what that paragraph is supposed to do and (in broad terms) how, then comment on critical pieces of code, e.g., note what a variable is supposed to contain. The analogous symbol in MATLAB is the percentage sign (%). Anything written after it is understood to be a comment.

Let us now add figure labels and formatting to the below code Fig. 3.6:

Python	MATLAB (suitable)
| ```
plt.ylim([0,10])
plt.title('this neuron spikes to repeated trials of
the same stimulus')
plt.xlabel('time (in milliseconds)')
plt.ylabel('trial number')
plt.yticks([x+0.5 for x in range(10)],[str(x+1) for x
in range(10)]) #1ᶜ
``` | ```
ylim([0.5 10.5])
title('this neuron spikes to repeated trials of the
same stimulus')
xlabel('time (in milliseconds)')
ylabel('Trial number')
set(gca,'Layer','top') %2
``` |

Align the yticks to be in the middle of each row, (*x*+1) sets first trial to 1. Set the label axis to the top so that green rectangle doesn't cover the axis.

Pseudocode
| ```
Create figure and specify subplot
Draw the stimulus presentation area green
For each of the ten trials
Extract the spike times from the spike train variables
Plot each row in the raster plot as vertical lines
Set the x-axis limits
Set the y-axis limits
Set the title
Set the x-axis label
Set the y-axis label
``` |

ᶜYou may encounter errors when typing between lines. The line of code: plt.yticks([x+0.5 for x in range(10)],[str(x+1) for x in range(10)]) can be typed across multiple lines in a text editor as:

plt.yticks([x+0.5 for x in \ range(10)],[str(x+1) for x in \ range(10)]).
```

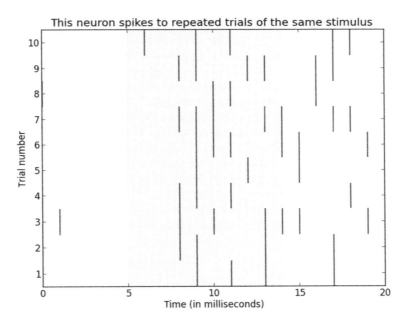

FIGURE 3.6 Raster plot of the neural response to 10 trials of a green light stimulus.

A raster plot yields a snapshot of *raw spike times* elicited by the stimulus across trials. We can see that the first value we calculated, the *first spike latency*, seems to vary considerably between trials: on some trials the first spike occurs at 8 ms, on some trials the first spike happens at 9 ms, on one trial it comes at 6, on a handful of trials it is 10, and on one trial there is a spike 4 ms before the light even comes on. We can also see that the neuron seems to discharge a spike a few times with each light stimulus, and it often also "fires" after the stimulus has turned off. There are many ways to characterize the spiking activity of neurons, some qualitative, some quantitative (like our example, *first spike latency*). Qualitatively, we ask ourselves if, once this neuron fires, does it keep firing? That is, is its ongoing activity tied to the stimulus?

There are many ways to quantify this, but it is a good habit to take a look at the raw data in the form of a raster plot to first get a qualitative sense of the spiking characteristics of the neuron. For instance, the *raster* plot in Fig. 3.6 seems to convey that the neuron responds to this particular light stimulus after 3–5 ms, and that its activity is maintained to some degree after the stimulus has turned off.

Does this falsify our hypothesis that the neuron always spikes 4 ms after the light turns on? It certainly looks like it—the spiking is not nearly as precise as we hoped it to be. But as we formed the hypothesis on a whim (based on the analysis of a single trial), we are free to change it. So let's say we hypothesize that the neuron fires *tonically at a rate of 500 spikes per second* to green light stimuli.

Let's unpack this statement. "Hypothesize that the neuron fires" is simple enough (the neuron discharges spikes) but we go on to make a prediction about the firing, that it fires "tonically." *Tonic firing* is spiking activity that is sustained or ongoing. This contrasts with *phasic* or *transient* firing, which is locked to the timing of the stimulus. We can come up with some metrics for quantifying the tonicity of the firing, but let's posit the qualitative hypothesis that it keeps firing and continue with the quantity "of 500 spikes per second." In our experiment, we didn't stimulate the neuron for an entire, continuous second, so we certainly won't have 500 spikes to count. However, the unit of spike rates is *spikes per second*, even if the instantaneous firing rate is sustained for much less than a second, just like you don't have to drive for an hour in order to go at 100 miles per hour at some point in your ride.

To know how many spikes we ought to expect in our short interval, we simply have to solve for x in the algebraic equation where x is proportional to *500 spikes* as our stimulus length (*9 ms—we really should have planned the experiment better in order to get easier math*) is to 1 s:

$$\frac{x \, spikes}{9 \, milliseconds} = \frac{500 \, spikes}{1000 \, milliseconds} \tag{3.1}$$

which, solving for x, leads us to expect 4.5 spikes for the duration of our 9 ms stimulus. The tail end of the stated hypothesis above was "to green light stimuli," which we sort of covered, and which we'll make more complex just when we start to get a better grasp of our results.

We thus need a way to visualize and condense the many stimulus trials and responses we recorded. We turn to the *peri-stimulus time histogram* (or *PSTH*) to visualize as a bar graph the spiking activity of the neuron over time, just before and after (hence peri) each stimulus. We also make use of multiple subplots within a single figure to compare the *rasters* to the *PSTH*. This is a standard depiction of neural data in early exploratory analysis and we'll revisit it in Chapter 4, Correlating Spike Trains.

An individual subplot is a subdivision of a figure. The code below indicates that the subplots will be arranged so that there are two rows and one column of subplots, and that we'll plot in the first of these. Note that though Python indexes lists by starting at 0, subplot indexing starts at 1 (!). If this seems inconsistent, it is because it is … inconsistent.

Pseudocode	Create the figure window using plt (originally imported via matplotlib) Create the first of 2 "subplots."
Python	```>>> fig=plt.figure()``` ```>>> ax=plt.subplot(2,1,1)```
MATLAB	```>> fig = figure;``` ```>> ax = subplot(2,1,1);```

We next create code for plotting (1) a subplot of spike rasters and (2) a PSTH based on that spiking activity.

We highly advise to comment so that (1) your future me can read and remember why you programmed something the way you did and what the variables stand for—which makes the code easier (or even possible) to maintain (2) so that other programmers can look at your code and have any chance of understanding what is going on.

In Fig. 3.6 we created the variable `spikeTimes` for each `trial` and plotted those values right away, overwriting `spikeTimes` with each new `trial`.

Pseudocode	Looping through *trial 0,1,...,9*
	    Get the index (time) of each spike and append to allSpikeTimes
	    Plot vertical lines for each trial
	Add the vertically spanning green box
	Set the limits of the y-axis to 0 and 10
	Add a title, a y-axis label, and an x-axis label to this subplot
	#1 Customize the labels of the yticks

Python	```python\nfig=plt.figure()\nax=plt.subplot(2,1,1)\nfor trial in range(len(tenSpikeTrains)):\n spikeTimes = [i for i,x in enumerate(tenSpikeTrains[trial]) if x==1]\n plt.vlines(spikeTimes,trial,trial+1)\nplt.axvspan(5,14,alpha=0.1,color='g')\nplt.ylim([0,10])\nplt.title('this neuron still spikes to repeated trials of the same stimulus')\nplt.xlabel('time (in milliseconds)')\nplt.ylabel('trial number')\nplt.yticks([x+0.5 for x in range(10)],[str(x) for x in range(10)]) #1\n```
MATLAB	```matlab\nax = subplot(2,1,1)\nrectangle('Position',[5,0,9,11],'FaceColor', [0.7 1 0.7],'linestyle','none')\nfor ii = 1:size(tenSpikeTrains,1)\n spikeTimes = find(tenSpikeTrains(ii,:)==1)-1\n line(repmat(spikeTimes,2,1),repmat([ii-0.5; ii+0.5],1,length(spikeTimes)),'color','k')\nend\nxlim([-0.5 20])\nylim([0.5 10.5])\ntitle('this neuron still spikes to repeated trials of the same stimulus')\nxlabel('time (in milliseconds)')\nylabel('Trial number')\nax.YTick = [0:1:10]\nset(gca,'Layer','top')\n```

We next sum across our `tenSpikeTrains` to see the total number of spikes that occur across all trials, using the function `bar` in POM. This function gives us a bar plot of the spiking as a function of time.

We also save the figure. Note the extension `.png` at the end of the string. We could also specify `.pdf` or `.jpg` or a few other image types.

Pseudocode	Now for the PSTH. We create our second subplot.
	Add the green background during stimulus time.
	Plot the bar plot #1 format bar (x-values, y-values, bar width)
	Add labels to the x- and y-axes of this subplot
	Save the figure
	Let's take a gander.

```python
>>> ax=plt.subplot(2,1,2)
>>> plt.axvspan(5,14,alpha=0.1,color='g')
>>> ax.bar(range(21),np.sum(tenSpikeTrains,0),1) #1
>>> plt.xlabel('time (in milliseconds)')
>>> plt.ylabel('# of spike occurrences at this time')
>>> plt.savefig('Figure with subplots of rasters and PSTH.png')
>>> plt.show()
```

```matlab
>> subplot(2,1,2)
>> rectangle('Position',[5,0,9,8],'FaceColor',[0.7 1 0.7]...
,'linestyle','none')
>> hold on
>> x=0:20;
>> bar(x,sum(tenSpikeTrains));
>> xlim([-0.5 20])
>> ylim([0 8])
>> xlabel('time (in milliseconds)')
>> ylabel('# of spikes counted at this time')
```

Putting it all together:

Python	MATLAB

```python
# The Python way for Figure 3.7.
fig=plt.figure()
ax=plt.subplot(211)
for trial in
range(len(tenSpikeTrains)):
spikeTimes = [i for i,x in enumerate(
tenSpikeTrains[trial]) if x==1]
plt.vlines(spikeTimes,trial,trial+1)
plt.axvspan(5,14,alpha=0.1,color='g')
plt.ylim([0,10])
plt.title('this neuron still spikes
to repeated trials of the same
stimulus')
plt.xlabel('time (in milliseconds)')
plt.ylabel('trial number')
plt.yticks([x+0.5 for x in
range(10)], [str(x+1) for x in
range(10)])
ax=plt.subplot(212)
plt.axvspan(5,14,alpha=0.1,color='g')
ax.bar(range(21),np.
sum(tenSpikeTrains,0),1)
plt.xlabel('time (in milliseconds)')
plt.ylabel('# of spike occurrences at
this time')
# You may get syntax errors
# when continuing onto the next line.
# Use the backslash character "\"
# in Python to allow a line break in
# the middle of code
# End Python code for Figure 3.7
```

```matlab
% What does the analogous Matlab code look like?
figure
subplot(2,1,1)
rectangle('Position',[5,0,9,11],'FaceColor',[0.7 1
0.7],'linestyle','none')
for ii = 1:size(tenSpikeTrains,1)
spikeTimes = find(tenSpikeTrains(ii,:)==1)-1
line(repmat(spikeTimes,2,1),repmat([ii-0.5; ii+0.5],1,length(spikeTimes)
),'color','k')
end
xlim([-0.5 20])
ylim([0.5 10.5])
title('this neuron still spikes to repeated trials of the same
stimulus')
xlabel('time (in milliseconds)')
ylabel('Trial number')
set(gca,'Layer','top')
subplot(2,1,2)
rectangle('Position',[5,0,9,8],'FaceColor',[0.7 1
0.7],'linestyle','none')
hold on
x=0:20;
bar(x,sum(tenSpikeTrains));
xlim([-0.5 20])
ylim([0 8])
xlabel('time (in milliseconds)')
ylabel('# of spikes counted at this time')
% End MATLAB code
```

Pseudocode

§ Begin English explanation of code for Fig. 3.7

Create the figure area

Specify that there are 2 rows and 1 column, and we'll start with the first

Plot the vertical ticks lines

Shade the area green to indicate the time of light stimulation

Set the lower and upper bounds of the y-axis

Set the title of the plot to 'this neuron still spikes to repeated trials of the same stimulus'

Set the x-axis label to 'time (in milliseconds)'

Set the y-axis label to 'trial number'

Set the y-axis tick locations and labels

Specify that we're making a subplot layout with 2 rows, 1 column, plotting in the 2nd row

Shade the stimulus area green

Make a histogram of all the spike times with bins from 0 to 20

Set the x-axis and y-axis labels

§ End English explanation of Python code for Fig. 3.7

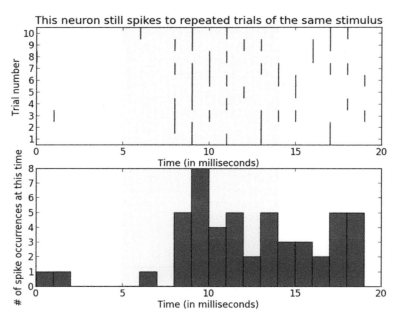

FIGURE 3.7   Raster plot (top panel) and PSTH (bottom panel) of neural response to visual stimulation in 10 trials.

Let's revisit our hypothesis, that the neuron fires *tonically at a rate of 500 spikes per second* to green light stimuli. We calculated that this would mean 4.5 spikes on average for the duration of our particular stimulus. But how can we infer this spike rate from the PSTH above? First, let us plot a new figure in which we scale the number of spike occurrences at each time by the number of stimulus trials (setting aside the rasters for a moment). This will show us, on average, how often the neuron spikes at each time point, and give us an estimate of the *spike probability* for each point in time.

We start by creating the figure, and omit the subplot line, noting that you don't need it for figures with single plots. Since we wanted to adjust the number of spikes for all the trials to form an estimate of *spike probability*, we will plot the mean of the spikes as a bar plot, instead of a sum.

If the neuron spiked, on average, at a rate of 500 spikes every second (every 1000 ms), then we might expect that for every millisecond there will on average be 0.5 spikes. Of course, we could not have performed this estimation with just a single trial as one cannot count half a spike. By measuring repeated trials, we form a more robust estimate of the spike rate over time, with the prediction (from our hypothesis) that the neuron will maintain a spike probability of 0.5 during the time that the stimulus is presented.

Let's draw a horizontal dashed black line (note the linestyle which can make it dotted or dashed, among other line types, and the *color = "k"* to de note *black*, we could have used *"r"* for red, or *"g"* for *green*, or *"b"* for *blue*) at the 0.5 spike probability threshold. Had we wanted a vertical line, we could have used the Python function plt.axvline.

Pseudocode	Create the figure Plot bar graph of the mean spikes Add horizontal line as a spike threshold
Python	```fig=plt.figure() plt.bar(range(21), np.mean(tenSpikeTrains,0),1) plt.axhline(y=0.5,xmin=0,xmax=20,linestyle='--',color='k')```

MATLAB	```figure
	bar(0:20,sum(tenSpikeTrains)./size(tenSpikeTrains,1));
	line(xlim,[0.5 0.5],'linestyle','--','color','k')```

Also label the axes and title, save the figure, and show it. Putting it all together, and using the simplified bar plot:

Python	MATLAB

<pre># The Python way for **Figure 3.8** 	
fig=plt.figure()
plt.axvspan(5,14,alpha=0.1,color='g')
plt.bar(range(21), np.mean(tenSpikeTrains,0),1)
plt.axhline(y=0.5,xmin=0,xmax=20,linestyle='--',
color='k')
plt.title('Spike probability given 10 stimulus
trials')
plt.xlabel('time (in milliseconds)')
plt.ylabel('probability of spike occurrences at
this time')
plt.savefig('**Figure 3.8**normalized PSTH with cutoff.
png')
# End Python code for **Figure 3.8**</pre> 	
	<pre>%What does the analogous Matlab code look
	like?
	figure
	rectangle('Position',[5,0,9,11],'FaceColor',
	[0.7 1 0.7],'linestyle','none')
	xlim([-0.5 20])
	ylim([0 1])
	hold on
	bar(0:20,sum(tenSpikeTrains)./
	size(tenSpikeTrains,1));
	line(xlim,[0.5
	0.5],'linestyle','--','color','k')
	title('Spike probability given 10 stimulus
	trials')
	xlabel('time (in milliseconds)')
	ylabel('probability of spiking at this
	time')
	% End MATLAB code</pre>

In MATLAB, we simply normalize by the number of spike trains. Again, we see the power of MATLAB when handling matrices. Thus, represent something as a matrix whenever possible, as it will allow to bring powerful tools to bear.

Pseudocode

§ Begin English explanation of code for Fig. 3.8

Create the figure plotting area

Shade the area green to indicate the time of light stimulation

Plot a bar plot of the mean of tenSpikeTrains

Plot a dashed black horizontal line at y = 0.5

Set the title to "spike probability given 10 stimulus trials"

Set the x-axis label to "time (in milliseconds)"

Set the y-axis label to "probability of spike occurrences at this time"

Save the figure to "Fig. 3.8 normalized PSTH with cutoff.png"

§ End English explanation of code for Fig. 3.8

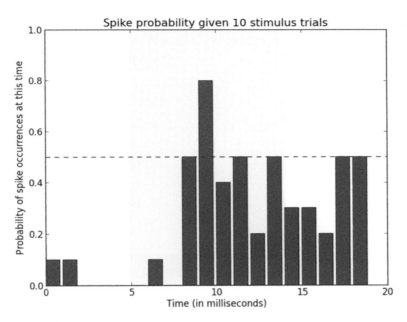

FIGURE 3.8   PSTH normalized by number of trials.

At first glance, it seems that our hypothesis of a consistent probability of 0.5 spikes over the stimulus interval does not hold and is false. We will discuss rigorous statistical tests at a later point. For now, we note that there are several phenomena that are not consistent with our hypothesis. For instance, there is a latency period before the spikes start, there is a phasic component of high spike probability around 4 ms and a tonically maintained probability around 0.4 or 0.5 for the remainder of the stimulus thereafter, even continuing for a while after the stimulus turns off. So it looks like things are more complicated than we initially expected, which (in biology) should be expected. To further illuminate what is going on, we could vary the intensity of the light, and hypothesize that the neuron fires more spikes with a shorter latency for brighter stimuli.

Let's load in the Python dictionary called `tenIntensities.pkl` (or in the case of MATLAB, the `.mat` file `tenIntensities.mat`), available for download via the companion website.

Python *dictionaries* are arranged in a manner where all of the *values* are assigned to a *key*. The *key* can be either a text string or a number, and the *value* can be almost anything: a number, string, list, array, or even another dictionary. To view the *keys* for this dictionary in Python , we type:

```
>>> tenIntensities.keys()
['4_intensity',
 '2_intensity',
 '8_intensity',
 '0_intensity',
 '7_intensity',
 '5_intensity',
 '9_intensity',
 '6_intensity',
 '3_intensity',
 '1_intensity']
```

Each key corresponds to the intensity of the stimulus, ranging from 0 to 9. So to get the *values* for, say, the *key* 4_intensity, we type:

```
>>> tenIntensities['4_intensity']
[[15.0, 15.0, 22.0, 25.0, 34.0, 23.0],
 [10.0, 32.0, 34.0, 22.0, 34.0],
 [13.0, 17.0],
 [9.0, 30.0, 36.0, 33.0],
 [8.0, 32.0, 31.0, 35.0, 19.0, 36.0, 19.0],
 [30.0, 13.0, 31.0, 36.0],
 [21.0, 31.0, 27.0, 30.0],
 [12.0, 15.0, 23.0, 39.0],
 [23.0, 30.0, 14.0, 23.0, 20.0, 23.0],
 [9.0, 16.0, 13.0, 27.0]]
```

We observe that each list within this value is another list of spike times. We use these spike times now to visualize the PSTHs over all stimuli.

Our raster and PSTH plotting techniques here are the same as before, with two main differences. The first, and most obvious from the output figure, is that we now have 20 subplots—10 rows and 2 columns. In the Python package matplotlib, the first subplot, referenced as *1*, is always at the top left. As we increase this index, our plot moves across each row to the right, to the end of the row, before moving down to the next column.

In the Python code, we make use of the `numpy` function `histogram`. It calculates the number of occurrences of values over a given range. It then returns the count of how many values occurred within each bin, and also returns the bins used, assigning these to the variables preceding the equals sign (and it doesn't plot anything). We use the variables obtained with `np.histogram` to make our histogram below with the function `bar`. In the code below we also give `numpy` a nickname, "np" which lets us refer to packages in a shorthand manner, e.g.: `np.histogram()`:

Python	MATLAB

```
# The Python way for Figure 3.9
import pickle
with open('tenIntensities.pkl', 'rb') as handle:
  tenIntensities = pickle.load(handle)
fig = plt.figure()
numIntensities = len(tenIntensities)
nbar={}
for key in tenIntensities.keys():
  ax=plt.subplot(numIntensities,2,float(key[0])*2+1)
  for trial in range(10): # this relies on there being 10
trials per stimulus intensity
plt.vlines(tenIntensities[key][trial],trial,trial+1)
  plt.xlim([0,20]);plt.ylim([0,10])
  plt.ylabel('intensity: '+str(key[0])+'\ntrial # ',style='it
alic',fontsize=5)
  plt.yticks(fontsize=5)
  plt.axvspan(5,14,alpha=0.1*float(key[0]),color='g')
  if float(key[0]) < 9:
    plt.xlabel('');plt.xticks([])
  else:
    plt.xlabel('time in milliseconds')
  if float(key[0]) == 0:
    plt.title('raster plot of spiking for each
intensity',fontsize=10)
  ax=plt.subplot(numIntensities,2,float(key[0])*2+2)
  plt.axvspan(5,14,alpha=0.1*float(key[0]),color='g')
  spikeTimes = [a for b in tenIntensities[key] for a in b] #1
  nOut,bins=np.histogram(spikeTimes,bins=range(20))
  nbar[float(key[0])] = nOut/10.
  plt.bar(bins[:-1],nOut/10.)
  plt.xlim([0,20]); plt.ylim([0,1])
  plt.yticks(fontsize=5)
  plt.ylabel('spike prob',style='italic',fontsize = 6)
  if float(key[0]) == 0:
    plt.title('PSTH for each intensity',fontsize=10)
  if float(key[0]) < numIntensities-1:
    plt.xlabel(''); plt.xticks([])
  else:
    plt.xlabel('time in milliseconds')
plt.savefig('Figure subplot 10 intensity rasters and psths.
png')
# End Python code for Figure 3.9
```

```
load('tenIntensities.mat')
figure
a= [1:2:20];
b =[2:2:20];
for ii = 1:size(A2,1)
subplot(10,2,a(ii))
if ii == 1
    title('raster plot for each intensity')
  end
rectangle('Position',[5,0,9,11],'FaceColor',
[1-(0.1.*ii) 1 1-(0.1.*ii)],'linestyle','none')
for jj = 1:10
spikeTimes = find(A2{ii,1}(jj,:)==1)-1
line(repmat(spikeTimes,2,1),repmat([jj-0.5;
jj+0.5],1,length(spikeTimes)),'color','k')
xlim([0 20])
ylim([0.5 10.5])
set(gca,'xtick',[])
end
end
%xlabel('time (in milliseconds)')
%ylabel('Trial number')
set(gca,'Layer','top')
for ii = 1:size(A2,1)
subplot(10,2,b(ii))
  if ii == 1
    title('PSTH for each intensity')
  end
rectangle('Position',[5,0,9,8],'FaceColor',
[0.7 1 0.7],'linestyle','none')
hold on
x=0:length(A2{ii,1})-1;
bar(x,sum(A2{ii,1}));
xlim([-0.5 20])
ylim([0 8])
set(gca,'xtick',[])
%xlabel('time (in milliseconds)')
%ylabel('# spikes')
end
```

Pseudocode

§ Begin English explanation of code for Fig. 3.9
Create new figure
Declare an empty dictionary *nbar*
For each key in the dictionary *tenIntensities*:

first column, raster plots for each intensity of the light stimulus. Plot in subplot corresponding to each intensity (an example *key* is: '*7_intensity*', so *key[0]* is the *0th* value of the string '*7_intensity*', which is '*7*', and *float('7')* equals 7.0, We take that value times two and add one since subplot indices count by row. For each trial this relies on there being 10 trials per stimulus intensity. Plot vertical lines corresponding to the spike times

Format the raster plots: Set the x- and y-axis limits. Set the y-axis label to the intensity, use '\n' as a carriage return, label the trial number, italic and fontsize, set the yticks fontsize. Add the green box, use the *alpha* value so that the transparency scales with intensity. If the intensity is less than 9, that is, if we are not plotting at the bottom. Do not label the x-axis.

Else, that is, if the intensity is 9. Label the x-axis as 'time in milliseconds'

If the intensity is 0, that is, if we are plotting at the top. Set the title to 'raster plot of spiking for each intensity' and font size to 10.

#1 perform list comprehension to unpack list

In the second column, plot the PSTHs for each intensity of the light stimulus. Plot the subplot in the second column, with each increasing intensity moving down a row. Plot the green box and set the *alpha* value to correspond to the intensity of the stimulus. Extract all the spike times for a stimulus intensity. Get *nOut*, a histogram array binned by the value *bins*. Add the values in *nOut/10.* to the dictionary *nbar* with key *float(key([0]))*. Plot the PSTH with *bar* function, calling all the bins except the last bin, and scaling *nOut* by the number of trials (10) .

Format the PSTHs: Set the x-axis and y-axis limits to [0,20] and [0, 1], respectively. Set the y-axis font size. Set the y-label to 'spike prob, make it *italic*, set the fontsize to 6. If we are in the first row (at the top). Set the title to 'PSTH for each intensity', with a fontsize of 10

If we are in any plot above the bottom plot, turn of the xlabel and xticks.

Else if we are at the bottom plot, set the xlabel to 'time in milliseconds'

Save the figure

§ End English explanation of code for Fig. 3.9

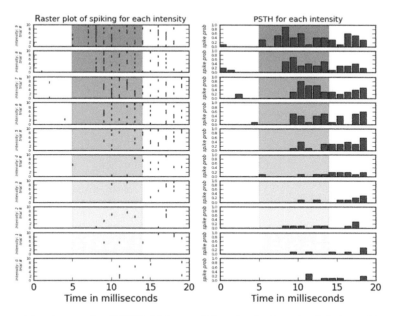

FIGURE 3.9 Raster plots (left panels) and normalized PSTHs (right panel) for stimuli of varying light intensities.

In the Python code, we slipped in the initialization of the *dictionary* nbar, which we did with curly braces {}. Later in the code, we assign *values* from nOut, which represent the number of occurrences of spikes for particular times, to *keys* of nbar. We access all the values of nbar below with nbar.values().

Our ability to measure *latency to first spike* here becomes quite difficult. We can qualitatively say that higher-intensity stimuli cause shorter latency responses. We will leave any swings at statistical inference to later chapters, and relish for now in our ability to make a colorful plots from the spike data.

Python	MATLAB
```# Begin the Python way for Figure 3.10 fig = plt.figure() ax = plt.subplot(111) aa = ax.imshow(nbar.values(),cmap='hot', interpolation='bilinear') plt.yticks([x for x in range(10)],[str(x) for x in range(10)[::-1]]) plt.ylabel('stimulus intensity') plt.xlabel('time in milliseconds') plt.title('heat map of mean spiking for various intensity stimuli') cb = fig.colorbar(aa,shrink=0.5) cb.ax.set_ylabel('mean spikes per time bin') # End Python way for Figure 3.10```	```for ii = 1:size(A2,1) A3(ii,:) = (sum(A2{ii,1})./10); end A3(:,22:100) = []; figure h = pcolor(A3); set(h,'Facecolor','interp') set(h,'Linestyle','none') set(gca,'YDir','reverse') colormap('hot') h = colorbar; ylabel(h, 'mean spikes per time bin') xlabel('time (in milliseconds)') ylabel('stimulus intensity') title('heat map of mean spiking for stimuli of varying intensity')```

Pseudocode

§ Begin English explanation of code for Fig. 3.10

Create plotting area

Specify that we're making one subplot

Plot the values of *nbar* as an image with a hot colormap and the colors bilinearly interpolated

Set where the y-ticks go and what their labels are

Set the y-*a*xis label to 'stimulus intensity'

Set the x-axis label to 'time in milliseconds'

Set the title to 'heat map of mean spiking for various intensity stimuli'

Create a colorbar, use the *shrink* command to customize its height

Set the colorbar label to 'mean spikes per time bin'

§ End English explanation of code for Fig. 3.10

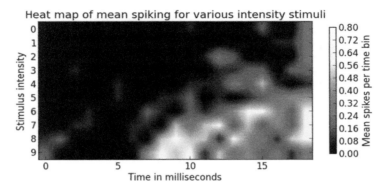

FIGURE 3.10   Heat map of mean spiking activity over time as a function of stimulus intensity.

Pseudocode	Create figure plotting area
	Specify two rows and two columns to plot, select the first one
	Plot the bar values as image with colormap hot, bilinear interpolation, and aspect 1.2
	Turn the y-axis and x-axis tick marks off
	Specify two rows and two columns to plot, select the second one
	Plot the bar values as image with colormap bone, nearest interpolation, and aspect 1.2
	Turn the y-axis and x-axis tick marks off
	Specify two rows and two columns to plot, select the third one
	Plot the bar values as image with colormap jet, bicubic interpolation, and aspect 1.2
	Turn the y-axis and x-axis tick marks off
	Specify two rows and two columns to plot, select the fourth one
	Plot the bar values as image with colormap cool, nearest interpolation, and aspect 1.2
	Turn the y-axis and x-axis tick marks off
	Save figure

(Continued)

| Python | ```
fig = plt.figure(); ax = plt.subplot(221)
aa = ax.imshow(nbar.values(),cmap='hot' ,interpolation='bilinear',aspect=1.2)
plt.yticks([]); plt.xticks([])
ax = plt.subplot(222)
aa = ax.imshow(nbar.values(),cmap='bone', interpolation='nearest',aspect=1.2)
plt.yticks([]); plt.xticks([])
ax = plt.subplot(223);
aa = ax.imshow(nbar.values(),cmap='jet', interpolation='bicubic',aspect=1.2)
plt.yticks([]); plt.xticks([])
ax = plt.subplot(224)
aa = ax.imshow(nbar.values(),cmap='cool',
interpolation='nearest',aspect=1.2)
plt.yticks([]); plt.xticks([])
plt.savefig('Figure 3.11- 4 heatmaps labels off.png')
``` |
| MATLAB | ```
figure
ax1 = subplot(2,2,1)
h = pcolor(A3);
set(h,'Facecolor','interp')
set(h,'Linestyle','none')
set(gca,'YDir','reverse')
colormap(ax1,'hot')
axis off
ax2 = subplot(2,2,2)
h = pcolor(A3);
set(h,'Linestyle','none')
set(gca,'YDir','reverse')
colormap(ax2,'bone')
axis off
ax3 = subplot(2,2,3)
h = pcolor(A3);
set(h,'Facecolor','interp')
set(h,'Linestyle','none')
set(gca,'YDir','reverse')
colormap(ax3,'jet')
axis off
ax4 = subplot(2,2,4)
h = pcolor(A3);
set(h,'Linestyle','none')
set(gca,'YDir','reverse')
colormap(ax4,'winter')
axis off
``` |

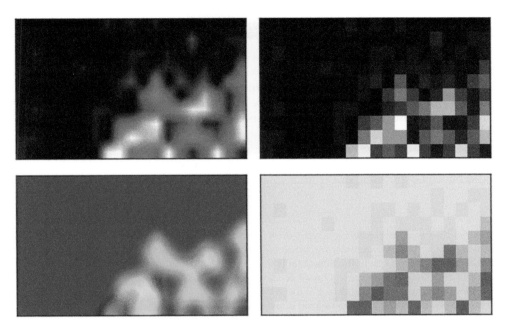

FIGURE 3.11   Heat maps with different color maps. Top left: "Hot", Top right: "Bone", Bottom left: "Jet", Bottom right: "Winter".

That's it for basic spike wrangling. Admittedly, this was somewhat of a toy example, but we have to get started somewhere. You'll find in your own dataset that you will need to employ a variety of wrangling methods to achieve your desired outcome—the goal here is to learn the various tools that help you to do so. In the very next chapter, Chapter 4, Correlating Spike Trains, we'll address a full-blown data analysis project, also involving spikes and their wrangling, but recorded from many neurons with many electrodes simultaneously.

QUESTIONS WE DID NOT ADDRESS

In the 10 intensity stimulus–response set, what order were the stimuli presented in? Did all of the bright intensity stimuli occur sequentially before moving to a lower intensity stimulus? If so, might the ordering of the stimuli influence the neuron's response characteristics? Might the neuron exhibit *adaptation*, where its firing probability adapts as a function of previous stimuli and its internal characteristics? How much time of no stimuli was between trials—what was the *inter-trial interval*, or ITI? How might changing the ITI influence the neuron's spiking?

What other optogenetic tools could we use to study the activity of single neurons? How many photons are required for the activation of ChR2? What sort of latency of activation patterns would we expect for other optogenetic tools? What sort of experimental preparation would be required to mimic the experiment in this chapter? What sort of experimental apparatus would be necessary to record from a single neuron?⌘

⌘. *Pensee on the proper unit of measurement for the spiking activity (firing rate) of single neurons*: Data result from the outcome of a measurement process. A unit of measurement is the fundamental unit which we use to express the quantity of a given quality. For instance, the currently agreed upon unit of measurement for length in the SI system is the meter, defined as "the length of the path travelled by light in vacuum during a time interval of $1/299792458$ of a second." (Taylor & Thompson, 2001). Consequently, all lengths that we wish to measure are then expressed in multiples of this reference length, e.g. 2 meters or 0.5 meters. This raises the question what the appropriate unit of measurement for a firing rate is. The firing rate of a neuron in response to a given stimulus (or even in the absence of a stimulus) is a quality of the neuron. The implication is that the neuron in question discharges action potentials (or "spikes") a certain number of times in a given interval, usually a second. The choice of this interval is probably what led to the fundamental confusion that one often sees in neuroscience publications. Firing rates are frequently expressed in terms of "Hz," e.g. "the neuron fired at a rate of 30 Hz." The "Hz" in question refers to the unit of measurement of a periodic, oscillatory process, namely 1 cycle (or period) per second. Unfortunately, this fundamentally mischaracterizes the very nature of action potentials. In contrast to harmonic oscillators, e.g., the motion of guitar strings, action potentials are neither cyclical nor periodic. Most frequently, they are conceptualized as "spikes," or point processes, in which case only the time when they occurred and how often this happened in a given interval is meaningfully interpretable. Spiking activity of single neurons is notoriously aperiodic and highly irregular—interspike intervals in a spike train are close to what would be expected from a Poisson process (Softy & Koch, 1993) and the variance of spike counts upon repeated stimulation suggests overdispersion (Taouali et al., 2016). Finally, it makes practical sense to avoid expressing firing rates in Hz in order to simply avoid the potential confusion when plotting it simultaneously with quantities that are appropriately expressed in Hz, such as the temporal frequency of a stimulus or the power of an analog signal in a certain frequency bin, as we'll encounter in Chapter 5. The debate about the theoretical significance of neural oscillations is heated enough (Shadlen & Movshon, 1999) without implying that spike rates are inherently oscillatory as well. But if not Hz, what is the proper unit of firing rate? As spikes are typically defined by the voltage trace recorded from an electrode in the brain crossing a reasonable threshold (and recorded as the time at which this crossing happened) then counted, it does make sense to express in units of impulses per second (ips) or spikes per second (sp/s) or simply events (threshold crossings) per second. All of these are conceptually sound and it is perhaps this range of equally suitable available options that prevented any of them from catching on as a consensus. In military contexts, the "rate of fire" (of rapid-firing guns) is typically expressed in rounds per second (rps), so by analogy, spikes per second (which is what we care about in a firing rate) is perhaps the most apt. Historically, there has been a movement to replace these with the eponym "Adrians," in honor of the pioneering Lord Edgar Douglas Adrian, the original discoverer of the rate code (Adrian, 1926) who won the Nobel Prize in 1932 and is the great-grandfather of many a neurophysiologist, if neurotree.org is to be believed. However, this unit did not catch on either—given the problematic nature of eponyms, this is perhaps just as well (Wallisch, 2011)—but almost anything would be better than expressing firing rates in Hz, which is fundamentally misleading. To repeat: a rate is not a frequency. For actual frequencies, the entire signal scales with it, if the frequency changes. In contrast, the individual action potentials remain invariant, regardless of spike rate. These are fast events, which have to be sampled frequently (or at high frequency) in order to be captured, even (or particularly) if the actual spike rate is very low. With this in mind, "spikes per second" really makes the most sense.

Correlating Spike Trains

We, the authors, vividly remember our first 5 years of coding with POM, specifically writing code to analyze data. Scientific programming is rather different from coding for say, building a graphical user interface (GUI). Everything gets more complicated when data get involved and we were admittedly rather lost during that entire time, writing convoluted code where programs included thousands of lines, code that was impossible to maintain or understand even after a short period of time. In essence, we were lacking principles of software development design for the analysis of data.

Since that time, we have hit on principles that work and we will outline and detail them in this chapter. It has not escaped our notice that these principles closely resemble what seems to have been implemented by the perceptual system (at least in the primate, to the degree of understanding that we have now). This makes sense: perceptual systems—we will show this with the example of the visual system because it is (to date) the most studied and perhaps best understood—are designed to analyze the environment for the extraction of relevant actionable information. They have been in development for hundreds of millions of years under relentless evolutionary pressure, yielding a high-performance analysis framework.

As far as we can tell, all sensory systems (with the exception of olfaction, which is special) follow the following five steps, and there are principled reasons for this (Fig. 4.1).

Step 1: Transduction. Every sensory system has to convert some kind of physical energy in the environment into the common currency of the brain. This currency is action potentials or spikes. In the case of vision, photons enter the eye through a narrow aperture (the pupil) and are focused on the retina by the lens. The retina transduces (i.e., converts) the physical energy of photons into action potentials. What leaves the eye is a train of action potentials (a spike train, see chapter: Wrangling Spikes Trains), carried by the optic nerve. The code equivalent of this is to write dedicated "loader" code. Its purpose is to load data from whatever format it was logged in—each physiological data recording

Neural Data Science.
DOI: http://dx.doi.org/10.1016/B978-0-12-804043-0.00004-0

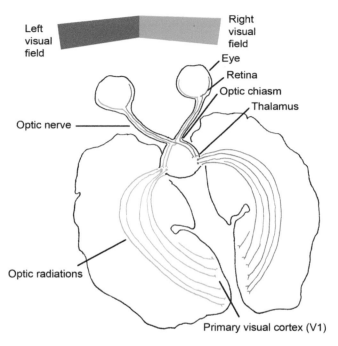

FIGURE 4.1 A cartoon of the primate visual system from the eye to V1. We omitted quite a few structures here, e.g., the superior colliculus or the suprachiasmatic nucleus, but this is the basic signal flow.

system creates its own data files, e.g., .plx files, .nev files, .nex files, .xls files, and .txt files are some popular formats. For you to be able to do anything with these in POM, they have to be converted into a format that POM can use first.

Step 2: Filtering. Once a spike train reaches the cortex, it will be processed. This can be a problem if it corresponds to information that is either not relevant at the time or if it is malformed in some way. In other words, the brain needs a gatekeeper to keep this irrelevant information out. In the brain, this step corresponds to the thalamus, specifically the lateral-geniculate nucleus in the visual system. This nucleus relays information from the retina to the visual cortex, but does so in a selective fashion. Of course this raises the issue of how the thalamus knows what irrelevant information is before it has been analyzed by the cortex. The brain solves this in several ways, including an analysis for low-level salient features, such as fast motion or high contrast, and then feeds that information back to the thalamus—there are many recurrent feedback loops to fine-tune the filtering process. We strongly recommend to implement this step in code, for similar reasons. It is very hard to write analysis code that is flexible enough to handle data that are not usable, be it due to parts of the data being missing, the study participant not doing what they are supposed to, the data being corrupted, the electrode not working, or the like. If such data enter the processing cascade, it usually renders the result meaningless or breaks the code altogether. It is best to avoid such data being processed in the first place. This step can be called "cleaning," "pruning," or "filtering" of the data. It is important that this step is performed agnostic to the results of the analysis. If you throw out data that don't conform to your hypothesis, you can get data supporting any hypothesis (this is data doctoring, which is unacceptable), don't do it. In contrast, this kind of integrity check before doing the full-blown analysis is critical. Another analogy for this step is that of a gatekeeper—the CEO (the cortex) can't be bothered with irrelevant information. There is too much information out there—if all of that was let in, nothing would get done. That's where a strict personal assistant comes in. Make sure to implement that in code.

Step 3: Formatting. The next step performed by the visual system is a categorical analysis of the data arriving from the thalamus. This step is performed by the early visual system, particularly V1. Here, the visual system determines the location and basic orientation of line segments (Hubel & Wiesel, 2004), and starts the process of determining what is foreground and what is background figure (Craft, Schütze, Niebur, & Von Der Heydt, 2007). Heuristically, this step can be understood as setting the stage for further analysis, not so much doing a lot of in depth analysis here already. The reasons for this will be understood more clearly when discussing the next step. Thus, we conceive of this step as "formatting" the data for further analysis. It is an absolutely critical step for data analysis. Once data are formatted properly, the rest of the analysis is usually rather straightforward. It might be unsettling to the beginner, but is not unusual to spend *most* of one's time writing analysis code in this step, simply "formatting" the data. Once data structures are set up properly, the actual analysis often corresponds to something very simple, like "loop through all participants in these conditions, then compare their means." Similarly, the visual system recognizes the importance of this step—the "early" visual system (V1 and V2) makes up about half the visual system by area in the primate (Wallisch, 2014).

Step 4: Computations. In the visual system, this step is implemented by extrastriate cortex—the cortical regions after striate cortex (or primary visual cortex) in the visual processing stream. Interestingly, whereas the previous steps have been done mostly in serial fashion, one after the other (the feedback to thalamus notwithstanding), this step is better referred to as steps (plural) because they happen in *parallel,* meaning that the signal might split into two or more copies so that multiple processes can occur on it simultaneously (Wallisch & Movshon, 2008). The fundamental reason for this is that many computations, in order to achieve the goal of the computation, have to abstract from some aspects of the source information, in effect destroying it. This might be information that is also important, but is better computed (in parallel) by another area. In effect, different parts of extrastriate cortex (at a minimum dorsal and ventral stream; Ungerleider & Mishkin, 1983) make copies of the information provided by V1 and work on that toward some outcome. For instance, in order to compute the speed of objects, it might be necessary to abstract from their location and identity, information that is also crucial to the organism, but can't be computed at the same time by the same area or serially. A parallel approach—working off copies of the original information—is perhaps the best way to solve this problem. We recommend to do something similar in code to implement this step. Specifically, we recommend to create as many parallel analysis streams as there are analysis goals. The number of analysis goals is given by how many theoretical questions need to be answered for any given project. For instance, it is conceivable that one analysis is concerned with the mean response of neurons under certain conditions whereas another deals with its variability—the underlying analyses are best done on copies of the original dataset and are complementary to each other. More analyses might build on these, e.g., a correlational analysis, as we will attempt here. We recommend to label these steps 4a, 4b, 4c, etc., in the code, signifying analysis steps that are in principle independent of each other, but can rely on each other, e.g., 4c being executed after 4a, etc. Note that *parallel processing* has a similar meaning in computer science, where computations are performed on data split onto different machines (Fig. 4.2).

Step 5: Output. This might come as a surprise to people living in the modern age, but the purpose of the visual system is not to provide fancy images for one's viewing pleasure, but to improve the survivability of the organism. This is true for sensory systems in general. Perception is not an end in itself—unless it results in motor output, the outcomes of its calculations are irrelevant. Over time, the system has been optimized to provide more adaptive outputs. In primates, the end result of the visual processing cascade in extrastriate areas is linked to motor cortex in order to transform the visual input to real-world outputs and memory systems to store information (the results of the computations on the inputs). We will do the same here, in code. Specifically, we will hook up the outputs of step 4, e.g., 4a, 4b to corresponding outputs, i.e., 5a, 5b. This analogy is tight. There are three principal kinds of outputs from the computation steps: Sometimes, we just want to output some numbers to the screen. Sometimes, the output will be a figure that visualizes the results of a computation. Usually, we also want to store the results so that we can load them in later without having to redo the entire analysis from scratch (this is particularly important as some analyses can take rather long). So just like the brain, we interface with long term memory systems at this point. In the case of code, this is the hard disk of your machine (or—rather soon—likely the cloud). You will want to have as many matching output functions (or files) as there are outputs of step 4.

And that's it. This is the general purpose framework ("the canonical data analysis cascade") that can be used for any data analysis project. We believe it is efficient and use it ourselves on a daily basis. As far as we can tell, most sensory systems do as well. Note that sometimes, some of these steps can be combined into one, e.g., we'll attempt to do the pruning/cleaning step at the same time as the formatting step in order to realize further efficiency gains. Sometimes you'll want to combine calculation and output steps, although one usually can output the same information in multiple

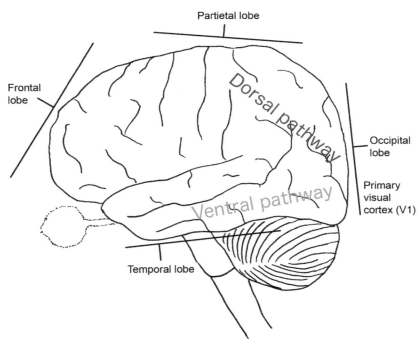

FIGURE 4.2 A cartoon of the extrastriate visual system beyond V1. It's endpoints interface with associative areas in the parietal lobe which are connected to motor systems in the frontal lobe and with memory systems in the temporal lobe.

ways. As long as you are careful when doing so, there is no problem with that, although we recommend separating the steps more strictly if you are an absolute novice until all of this becomes second nature.

To conclude, we strongly advise to recreate these five principal steps in code, when attempting any full-scale data analysis project. Specifically, we recommend to partition analysis code into these steps, i.e., either by writing individual files that correspond to each step or by dividing the code up into self-contained segments. What is important is that each logical segment fits on a screen (or just about fits), if it does not, it will be very hard to maintain. In our experience, analysis code has to be revisited with surprising regularity (e.g., when reviewer 2 asks for additional analyses) and unfortunately, memory for code seems to be particularly ephemeral transient. In other words, if you do not organize and comment your code well, you will not understand the code you wrote, even after a surprisingly short time interval. This can put you in a difficult position, for instance when pressed for time, as in looming grant or conference deadlines. Avoid it.

In addition to these five steps implemented by sensory systems of the brain, we recommend adding a zeroth and a sixth step. The zeroth step is an "initialization" step (in the brain, this might correspond to birth) you want to start from as nearly blank a slate as possible. In our experience, many logical errors in programming are due to something still lingering in memory that you forgot about but that affects the execution of your code. These errors are hard to track down and can be catastrophic. It is best to avoid them and the best way to do that is to clear the memory before doing anything else. In addition, you will want to define some constants, parameters, or analysis flags in this step. The reason why you want to do this here (and not strewn throughout the code) is that you have an easily accessible section—all in one place, at the beginning of the code—that governs the execution of the code, e.g., options to run the analysis with or without normalization, setting which baseline to use and things like that. Even the brain (at birth) doesn't start with a completely blank slate (due to the complex nature of the development of synaptic organization via both genetics and maternal environment). There is a reason for that.

Finally, in the spirit of making the code maintainable, and even runable a couple of months after writing it, we recommend writing a file that corresponds to the sixth step, which is kind of a "wrapper" and "readme" file that (if you wrote a file for each of the five steps) calls the right files with the right parameters in the right order and some kind of documentation of what is going on. This superstructure perhaps corresponds to the brain as a whole. Although we haven't managed to find the documentation yet.

Without further ado, let's jump into a full-blown data analysis project. Specifically, we will recreate some of the analyses from Snyder, Morais, Willis, and Smith (2015) from scratch. This paper attempts to answer the question whether the spiking of individual neurons is influenced by the activity of the network that they are embedded in. To do so, Snyder et al. recorded signals from a visual area with a 96 electrode array, splitting the signal from each electrode into a spike channel and a channel of analog signals, or local field potentials (LFPs) while the animals were shown oriented gratings. For the purposes of this chapter, we focus on the information from the spike channels. We will address the analysis of analog signals in frequency space in Chapter 5, Analog Signals.

Thus, we will use 96 time series of spiking information as source data in this chapter. We already encountered one time series in Chapter 2, From 0 to 0.01 and Chapter 3, Wrangling Spikes Trains. As a general rule, much of the data in neuroscience will come in the form of time series. Prominent examples would be EEG, MEG, EMG, LFP, spike trains, participant responses, and the like, so knowing how to handle them is important.

A time series is a vector of system states (e.g., the voltage around an electrode tip), where successive elements in the vector represent successive time samples at which the state of the system was measured.

As we have more than one time series (96 in fact) we can now look at the relationship between time series from different electrodes in order to calculate a quantity that is called the "spike count correlation." We will explain the concept of spike count correlations and why they matter in detail later, but let's first build up to that. It is time to get started.

In previous chapters, you could input all code from the command line. This project is too big for you to do so. We strongly urge you to start a new script in the code editor in MATLAB and a notebook in Python.

The first thing you'll want to write, for any POM file, is a header. The header indicates, as concisely as possible, what the program is and what it does, usually by stating its inputs, output, and assumptions. You'll also want to include the person responsible for coding this and how they can be reached (in case there are any questions). Finally, we recommend to include information about version history. As all of this information is meant to be read by you (not POM) it needs to be "commented out." We are here showing the MATLAB version of a typical header file for this program. In MATLAB, comments are indicated by % and rendered by default in green. So for the Python version, replace all % with # and expect it to be rendered in gray or light blue.

```
%*****************************************************************************
%This program recreates the spike count correlation analysis from Snyder et al. (2015)
%It assumes that the data file from the recording array exists in the same
%directory as the analysis file.
%Name and email of code creator
%V1: 06/16/2016: Steps 0-3
%V2: 06/20/2016: Steps 4 and 5
%V3: 06/21/2016: Adding parameters to step 0
%*****************************************************************************
```

For the purposes of this chapter, we will create a single file that includes all steps of our canonical data analysis cascade in sequence. We could have written separate files or functions for each of them, but will not do so at this time. Instead, we separate these steps into code segments, "sections" in MATLAB and "cells" (not to be confused with neurons) in Python. In MATLAB, code sections are separated by double percentage symbols %%. We recommend writing a brief paragraph as to what the section is supposed to accomplish at its beginning.

So we start with step 0: Initialization. You can write the paragraph explaining what the section does in pseudocode. We will do this once explicitly, for this section, in MATLAB format. For other sections, we will just implement the Rosetta stone approach and rely on you to understand where to put the pseudocode (in comments after the section header).

```
%% 0: MATLAB Init
%In this step, we clear the memory, close all figures and clear the
%screen in order to start from a blank slate.
%We then define constants that we will use later.
```

```
## 0. Python Init
# For Python, we load the functions that we'll be using and assign
# some of them nicknames
```

Python	MATLAB
import scipy.io	clear all
from collections import defaultdict	close all
import matplotlib.pyplot as plt	clc
import numpy as np	numChannels = 96;
import scipy.stats as sc	noiseCodes = [0, 255];
numChannels = 96	timeBase = linspace(-0.2,2.5,270);
noiseCodes = [0,255]	gratingOn = 0;
timeBase=np.arange(-.2,2.5,.01)	gratingOff = 2;
gratingOn = 0	
gratingOff = 2	

Constants can be anything, often they are parameters that modify the execution of the analysis code that follows. Here, we specify that we expect to analyze data from 96 channels (electrodes). We could have gotten that information from the data file itself, but we haven't loaded it yet and we do have prior knowledge about the experiment from the paper, so we might as well use it here. This also exemplifies the power of setting parameters early in the file. For instance, say we receive data from a similar experiment, but one that used a 256 electrode array instead. We can easily modify the code here in order to accommodate the analysis from the larger array. The variable noiseCodes represents spikes that we consider to be noise. This is also information that we received from the originator of the data—if we want to change what we consider noise, e.g., to run a more permissive or a more conservative analysis, we could easily do so here, by changing the noise codes. More on noise codes shortly. Finally, for now, we assign the variable timeBase as a linear vector going from –0.2 to 2.5 in 270 steps of 0.01. We will use this variable as time base. We learned from the paper that in each trial, the grating comes on 150 ms after fixation and stays on for 2 s. In other words, this time base should be sufficient to see the rise and fall of spiking activity in response to the grating stimulus and show quite a bit of the baseline as well. A bin width of 10 ms seems reasonable. Should we want to change any of these, e.g., showing more of the baseline or desiring a different temporal resolution, we can easily do so by changing timeBase here. We also input other knowledge we have about the structure of the experiment, e.g., that the grating came on at 0 s and that it was extinguished from the screen at 2 s, for any given trial. We use this knowledge about the trial structure much later, when we plot time courses. Note the close correspondence to organisms. We clear the memory to start with a blank slate, then repopulate the workspace with our knowledge about the world—the nature of the data we want to process—in the case of an organism, it would be the kind of information that the organism is likely to encounter. Such is the power of priors and of the canonical data analysis cascade—for both organisms and data analysis programs.

So let's move on to step 1, the loader:

Pseudocode
1: Loader. Load the data file

Python	MATLAB
matIn=scipy.io.loadmat('arrayDATA.mat')	load('arrayDATA.mat');

To make your life easier, we already preprocessed the file, so that it contains a single structure that contains all the data that are relevant here, and it is already in MATLAB format (as indicated by the .mat file).

That structure is called DATA. You can glean its properties by typing DATA in MATLAB (or matIn['DATA'] in Python. If you do so, you'll see that it contains 2300 rows with two fields: nev and ori. Each row represents a trial. ori represents the orientation of the grating for a given trial and nev is a matrix that represents the spiking information. Each trial contains its own nev matrix. Rows in the nev matrix represent individual spikes per trial, one for each spike during the trial. Column 1 represents the electrode on which the spike was detected, column 3 represents the corresponding time since the beginning of the trial at which the spike was detected, and column 2 represents a "sort code." The sort code is a number from 0 to 255. Neural data, in particular data from electrode arrays, have to be carefully sorted before they can be processed further. The reason for this is that each electrode records the voltage at the tip at a given time, but this voltage is the result of the spiking activity of an unknown number of neurons in an unknown configuration around the electrode tip (Lewicki, 1998). Thus, spike sorting—resolving the individual sources that all contribute to the same voltage sum—is fundamentally an inverse problem and remains rather tricky, despite best attempts at solving it (Pillow, Shlens, Chichilnisky, & Simoncelli, 2013). Here, take solace in the fact that spike sorting has already been done, and done carefully. This is indicated by the sort codes assigned to each spike. Sort codes 0 and 255 represent noise (accidental threshold crossings or threshold crossings for artificial reasons such as animal movement or instrument noise) and will need to be discarded from further analysis. Other sort codes –1 to 254 represent units (where one unit is thought to be nearly one neuron) that were assigned to different clusters. If a spike has the same sort code, it was discharged from the same neural cluster (either multiunit or the same single unit).

Structures are nested (or hierarchical) and labeled matrices. Each field has a label (such as ori) and a content (could be matrices of numbers or characters). Structures can contain arbitrary numbers of levels. Structures make for great storage units, but doing computations on them can be challenging, which is why we'll need further processing steps (particularly step 3, formatting) of our canonical data analysis cascade. Fields are accessed by typing a period or dot. If you don't remember the field names, you can tab-complete them. For instance, to access the orientation of the 57th trial in MATLAB, type DATA(57).ori—the output indicates that the orientation of the grating was zero degrees for that trial. If you want to know the spiking activity from trial 210, type DATA(210).nev to reveal that a brisk 1747 action potentials from a wide range of channels were discharged during that trial.

Other parameters that we will use in the analysis, such as the number of trials or the number and type of orientations involved, we get directly from the data themselves, like this:

Pseudocode

Determine several parameters that will be used in later analysis from the data
itself, such as the number of trials, the grating
orientation and number of grating orientations.

Python	MATLAB

```
numTrials = len(matIn['DATA'])              numTrials = length(DATA);
allOris=[matIn['DATA'][_][0][1][0][0] for _ in\   ori = unique([DATA(:).ori]);
range(numTrials)] #intermediate variable    numOri = length(ori);
ori = list(set(allOris))                    trialIndices = cell(1,numOri);
numOri = len(ori)
trialIndices = defaultdict(list)
```

---

numTrials represents the number of trials and can be gotten simply by counting the number of rows of the DATA structure. The variable ori represents the number of unique grating orientations used in the experiment. Here, it was two orientations, 0 and 90, but we want to write the code so that it is flexible enough to accommodate different orientations (and different numbers of orientations) in the future, should we receive them. To do so, we use the function unique in MATLAB, and set in Python. Each returns an ordered list of all unique orientation values. We provide this function with a vector of all orientations from the structure, as you can see in the code above.

trialIndices will contain the trial indices in the DATA structure that correspond to a given grating orientation. Note that we allocate trialIndices as a cell array in MATLAB, and as a defaultdict (a special kind of dictionary that allows the user to append to values even if the keys do not yet exist) in Python. The number of columns of this cell array corresponds to the number of unique orientations. The reason we want to allocate it as a dictionary or cell array in POM is because we want to allow for the possibility that not each orientation was presented an equal number of times (although this is the case here). For instance, if one orientation was presented 1200 times and the other 1300 times, and we represent this as resident matrices of a cell array, this is no problem. If we represented it as a 1150 × 2 matrix (one row for each trial, different orientations in each column), we would have to assume that each orientation was presented exactly the same number of times. If this was off by a single number, we could not represent it as a matrix. This is an unacceptable degree of rigidity, so in order to make our code more flexible, we represent these indices as a dictionary or cell array from the beginning. Or rather, we will do so. For now, we are just preallocating them by declaring their structure. If you type `trialIndices`, you will see that there is nothing in the matrices, but the structure of the cell array itself—a 1×2 matrix of matrices has been established in MATLAB, and an empty dictionary in Python: `defaultdict(list, {})`. It is good practice to preallocate these things in memory before using them, for a variety of reasons. More on this shortly.

It is now time to put the data into a format that will enable us to do the computations we need to do. We also need to identify (and remove) traces that crossed the spike threshold accidentally, but are actually noise, so as to not pollute the analysis. Here, noise is our primary concern. In other experiments, we might have to deal with missing data or data that are not usable for other reasons (of which there are many). For reasons of efficiency, we will do these two steps in one, here (we could loop through the data twice) first identifying the valid data, then reformatting it, but we might as well do it in the same step. Note that we are not vectorizing the code here. To more easily see what is going on, we use nested loops. That is not the most efficient way of doing things, but given that this is a relatively manageable dataset, acceptable. We can vectorize this code later. Finally, all of this is known under the moniker "data munging" in data science itself.

Step 2 + 3: Formatting of valid data (which implies the identification and elimination of invalid data)

Pseudocode	2+3: Pruning and Formatting Step 1: Identify trial numbers (indices) that correspond to a given orientation

Python	```for tt in range(numTrials):```
	``` for oo in ori:```
	``` if allOris[tt] == oo:```
	``` trialIndices[oo].append(tt)```

MATLAB	```for tt = 1:numTrials```
	``` for oo = 1:numOri```
	``` if DATA(tt).ori == ori(oo)```
	``` trialIndices{1,oo} = cat(1,trialIndices{1,oo},tt);```
	``` end```
	``` end```
	```end```

There is a lot going on here, so let's unpack it one by one. First of all, note that we use nested loops. First, we go through all the trials from 1 to 2300. Within that loop, we first determine what is in the orientation field of the DATA structure for that trial and put it into a temporary variable. We then go through the number of orientations (here two) and see whether the orientation we are checking for corresponds to the stimulus orientation for that particular trial. We do so via using the function if, which is a control structure. if is a popular way to fork the code, to do something if (and only if) the conditions following the if statement are met (potentially do something else if it is not met, but here we only indicate what is to be done if the conditions are met). Specifically, what we are asking it to do is that if the orientation of the given trial is equal to the first orientation, put it in the first column of trialIndices, if the second orientation matches, put it in the second column, and so on. We end up with a list of trial indices (trial numbers) that match the respective orientation in the variable trialIndices. We do so by concatenating the existing contents of the respective matrix in the trialIndices cell array and the new trial number, by using the function cat in MATLAB. This function concatenates along the nth dimension (here, we indicate 1, for rows), the existing piece plus the new piece. In Python, we make use of the append function, which we already saw in Chapter 3, Wrangling Spikes Trains.

This is why preallocation was important—this needs to work for the very first trial, when we concatenate the first trial number with an empty (but existing) matrix.

Also note that we use several distinct counters in this nested loop, namely tt and oo. We strongly recommend to use meaningful indices, such as tt for trials and oo for orientations. oo is technically not optimal because it could at first glance be confused with two zeroes or the infinity sign ∞; perhaps we should have used rr for oRientations. Be creative. Mathematicians and engineers are fond of using the generic i, j, and k. We don't recommend doing that. If you write a nested loop with these, you are bound to forget what is looping over what, quickly.

Note that we used smart indent for MATLAB as well, but didn't have to. Loops and if-statements are closed with corresponding end statements. In Python, we do not close the loops with end statements, as they are implied by the blankspace.

Finally, note that we didn't have to use the temporary variable temp, either. We discourage the use of temp because it slows down the code and is rarely needed. We simply put it here for didactic reasons, to explicitly unpack what is going on into logical steps. You can remove it if you restate the line

```
if temp == ori(oo) as if DATA(tt).ori == ori(oo).
```

You will end up a list of trial numbers with trials that correspond to a given orientation (per column) either way. Try it; it is important to do manipulation checks. Never assume that your code worked. Always check with some suitable chosen test cases that it does what you think it does. Neglecting to do so has sunk papers. This is a big deal—you don't want to end up on retractionwatch.com.

For instance, you could pick a few trial numbers from the first column of trialIndices and make sure that they all correspond to the same orientation in DATA, and a few from the second column of trialIndices that they all correspond to the other orientation in DATA. To do this, you'll have to reach into the trialIndices structure, which is a cell array. Which is a good time to introduce the third, and final, kind of parentheses in MATLAB, namely curly braces— { }. Using regular parentheses with trialIndices will simply return how *many* elements it contains and in what arrangement, e.g., trialIndices(1,1) will report that it contains a 1150 × 1 matrix. To actually access the contents of trialIndices, you'll have to type trialIndices{1,1} for the first column and trialIndices{1,2} for the second column.

Now that we know which orientation was run in which trial, we can start reformatting the data to suit our analysis in earnest. Specifically, we want to find the spikes that were discharged during any given trial. In this case, due to the way the raw data were formatted, it makes sense to do step 2 (data cleaning) and step 3 (data formatting) at once. In other words, identify only valid spikes. For other experiments, e.g., if there is a suspicion that some of the trials are not good, it would make sense to identify bad trials before looking for the trial numbers and have a dedicated cleaning stage. As we will implement our formatting and cleaning as a nested loop with three levels of hierarchy, it makes sense that we do everything at once, for efficiency's sake.[¶]

We'll then walk you through it. But note that this is deceptive. When writing such code, we *strongly* advise to code loops *"from the inside out,"* e.g., put in some placeholders for the indices (your test cases) to make sure that the code does what you think it does for one instance in the innermost loop, here finding spikes from *one* channel, for *one* grating orientation and *one* trial, then repeat this with some other test cases, then put a loop around it, then a loop around that and so on, until you have achieved all the looping that you want or need to do.

In this spirit, in the pseudocode below, we will describe what happens within the innermost loop first, then what we are looping over. We indicate where we are in the code with a number that is commented out at the end of the command. These *"comment pointers"* are useful when commenting your own code—it will make the logic that much more explicit and the code that much more maintainable.

The code below assumes an equal number of trials per orientation, which is the case here. Without further ado:

Pseudocode	
	1. Find the indices of spikes that were recorded on a given electrode (channel), from a given trial and condition (grating orientation)
	2. Use those indices to find spikes that were flagged as one kind of noise
	3. Use them again to find spikes that were flagged as another kind of noise
	4. Merge the two sets of noise into a big "killset"
	5. Eliminate all indices that correspond to noise from the variable that contains spike indices for a given electrode (channel), trial, and condition
	6. Find the corresponding spike times for the remaining valid spike indices and put them in the large storage structure, the spikeTimes cell array
	7. Loop through all channels (electrodes)
	8. Loop through all conditions (grating orientation types)
	9. Look through all trials of a given type

(Continued)

| MATLAB | ```
spikeTimes = cell(length(trialIndices{1,1}), numOri, numChannels);
for tt = 1:length(trialIndices{1,1}) %9
for oo = 1:numOri %8
for cc = 1:numChannels %7

tempSpikeIndices = find(DATA(trialIndices{1,oo}(tt)).nev(:,1) == cc); %1
kill1 = find(DATA(trialIndices{1,oo}(tt)).nev(tempSpikeIndices,2)==noiseCodes(1)); %2
kill2 = find(DATA(trialIndices{1,oo}(tt)).nev(tempSpikeIndices,2)==noiseCodes(2)); %3
killSet = union(kill1,kill2); %4
tempSpikeIndices(killSet,:) = []; %5
spikeTimes{tt,oo,cc} = DATA(trialIndices{1,oo}(tt)).nev(tempSpikeIndices,3); %6

end %7
end %8
end %9
``` |
|---|---|
| Python (near equivalent) | ```
numOris = len(ori)
allOris=[matIn['DATA'][_][0][1][0][0] for _ in\
range(numTrials)]
for orind,oo in enumerate(ori): # %8
    for tt in trialIndices[oo]:# %9
        for cc in range(numChannels): #%7
            spikeTimes[tt][orind][cc] = [spike[2] for spike in
matIn['DATA'][tt][0][0] if (spike[0]==cc) and (spike[1] not in noiseCodes)]
``` |

In essence, this code populates the cell array spikeTimes, which will contain the valid spike times for a given trial, grating orientation and channel. We recommend to preallocate it in memory in MATLAB by adding this line before the loop:

```
spikeTimes = cell(length(trialIndices{1,1}), numOri, numChannels);
```

In Python we initialize spikeTimes through a nested list comprehension, where we build out empty lists [] where there are a numChannel number of []s inside a numOri number of []s inside a numTrials number of []s. Note that this can be prohibitively slow with these loops, so we offer an alternative below.

We could have used other indices, e.g., condition instead of orientation and electrode instead of channel. It does not matter what you call it as long as you keep it consistent.

In the spirit of everything else in this chapter so far, this version of the code is rather explicit and step by step. There are many other ways of doing this, most of them more succinct. For instance, instead of first creating a temporary spike times variable, then identifying various types of noise, then combining those, then eliminating the noise, then assigning what is left to `spikeTimes`, we could have simply done all of that in one (intense) line, by identifying the spike times of valid trials in one go, replacing steps 1–6 with this single line (separated by ellipses for clarity), basically a string of nested logicals that have to be true at the same time (trial indices that match a given channel, but not any of the noise codes), using some of the components of lines 1–6 from above to achieve the same outcome:

```
spikeTimes{tt,oo,cc} = DATA(trialIndices{1,oo}(tt)).nev(...
   intersect(find(ismember(DATA(trialIndices{1,oo}(tt)).nev(:,2),noise) == 0),...
      find(DATA(trialIndices{1,oo}(tt)).nev(:,1) == cc)),3);
```

In essence, this "line" of code finds, for any combination of trial number and grating orientation, the spikes at the intersection of a given channel that are not a member of the noise set, gets their spike times, and assigns them to the appropriate place in the cell array.

As you can see, this code is much more compact (one command line instead of six), but also much harder to understand and maintain. This is a fundamental programming tradeoff in general. What balancing point in this trade-off you are most comfortable with will depend on your experience level and coding goal.

You can convince yourself by trying some examples that this line produces the same output as the explicit steps from 1 to 6. Which method is more elegant? That depends on you.

As this example demonstrates, you can always write more compact and "efficient" code, but it also illustrates why you might not want to.

The final formatting step that we'll do for now before moving on to do some computations and plotting concerns abstracting from trial numbers. Right now, the cell array `spikeTimes` contains valid spike times for a given combination of trial number, condition, and electrode. This will be important later, but first we'll want to look at *all* spikes from a given electrode and condition to ascertain which channels are even viable and whether the condition made any difference.

But our work is not in vain. We can now loop through the `spikeTimes` cell array to abstract over trial numbers. Before we do so, you should preallocate the variable `linearizedSpikeTimes`, another cell array by adding the following line to the script:

```
linearizedSpikeTimes = cell(numChannels,numOri);
```

Pseudocode	1. Find all spike times for a given combination of trial, orientation and channel in the cell array spikeTimes and add it—by concatenating—to the appropriate place in the cell array linearizedSpikeTimes
	2. Loop over all trials
	3. Loop over all orientation conditions
	4. Loop over all channels

Python

```
linearizedSpikeTimes = defaultdict(lambda: defaultdict(list))
for eachtrial in matIn["DATA"]: #2
    stimori = eachtrial[0][1][0][0]
    for eachspike in eachtrial[0][0]:
        if eachspike[1] not in [0,255]: # if the value is good
            trode = eachspike[0] #4
            spikeTimes = eachspike[2]
            linearizedSpikeTimes[trode][stimori].append(spikeTimes)
```

MATLAB

```
linearizedSpikeTimes = cell(numChannels,numOri);
for cc = 1:numChannels %4
 for oo = 1:numOri %3
  for tt = 1:size(spikeTimes,1)%2
   linearizedSpikeTimes{cc,oo} = …
   cat(1, linearizedSpikeTimes{cc,oo}, spikeTimes{tt,oo,cc}); %1
  end %2
 end %3
end %4
```

The end result of this is a cell array linearizedSpikeTimes that contains one long column with valid spikeTimes per combination of electrode and condition, with as many rows as there were spikes. Try a couple of specific combinations to convince yourself that this is in fact the case.

We are now in a position to do the first calculation and plotting on the basis of this formatted data. The data structure is general enough to allow several other computations that we'll do later. We call such a combination of computation and plotting a "computation and plotting stream," within the canonical data analysis cascade.

The first such stream, 4a/5a computes the peristimulus time histograms (PSTHs) for each channel and condition and graphs them in a data browser.

Step 4a: Calculating PSTHs: Before executing the code below, preallocate the matrix PSTHs with the following line. This will be a three-dimensional matrix, not a cell array, as each number of rows and columns per sheet is the same. We initialize with zeros by calling the function *zeros*:

```
PSTHs = zeros(numChannels,numOri,length(timeBase));
```

In Python, instead of preallocating these values to an array of zeros (which we could have done, and encourage the reader to try), we make use of the defaultdict function, which allows us to dynamically add *key:value* pairs to the dictionary PSTHs. However, we make use of the special nested function using lambda, which allows an additional layer of dictionaries. The *values* for the first level of the dictionaries are in fact *keys* for the next level down. This technique is useful if you have specialized indices or *keys* you wish to keep organized, which we have in the case of stimulus orientations and channels.

Pseudocode	1. Parse the suitable position in the cell array linearizedSpikeTimes by the edges contained in timeBase. Count how many spikes fall into each bin and assign that to the appropriate location in the 3D-matrix PSTHs
	2. Loop through all conditions
	3. Loop through all channels
| Python | ```
PSTHs = defaultdict(lambda: defaultdict(list))
for unitkey in linearizedSpikeTimes.keys():
    PSTHs[unitkey][0].bins = np.histogram(linearizedSpikeTimes[unitkey][0], bins=timeBase)
    PSTHs[unitkey][90].bins = np.histogram(linearizedSpikeTimes[unitkey][90], bins=timeBase)
``` |
| MATLAB | ```
PSTHs = zeros(numChannels,numOri,length(timeBase));
for cc = 1:numChannels %3
 for oo = 1:numOri %2
  PSTHs(cc,oo,:) = histc(linearizedSpikeTimes{cc,oo},timeBase); %1
 end %2
end %3
``` |

In MATLAB, we use the function histc to parse the long list of spike times for any given combination of electrode and condition. histc returns a vector of bin counts. We supply the parameter timeBase that we defined in the initialization as bin edges.

Now we immediately follow 4a by the way we want to plot it, in the form of a data browser. The idea is that the user will be able to see all spikes from a given channel and condition over time (the peristimulus time histogram), with markers of stimulus on- and offset, and being able to loop through the data in this way.

5a: The data browser. Note that you can also write this piecemeal in order to make sure that everything works. For instance, all the bells and whistles (axis labels, etc.) are not needed until the end.

Pseudocode	
	1. Open a new figure
	2. Make the background white
	3. Plot the PSTH of a given channel and orientation over the timebase
	4. Indicate stimulus on- and offset through black vertical lines
	5. Plot only from beginning to end of timebase
	6. Add x- and y-labels
	7. Add a title that indicates which channel and condition we plot. Use unicode to depict the degree symbol
	8. Make font size of labels bigger
	9. Change position of ylabel so as to not collide with numbers (in data units)
	10. Loop through orientation conditions - two per figure
	11. Show graph and wait for user input (keypress) before moving on
	12. Loop through channels

Python

```python
for unitkey in PSTHs.keys():
    fig=plt.figure(facecolor='w')
    for orind,oo in enumerate(ori):
        ax = fig.add_subplot(2,1,orind+1)
        ax.plot(timeBase[:-1],PSTHs[unitkey][oo],lw=3,color='b')
        ax.set_xlim([-.2,2.5])
        ax.vlines(gratingOn,0,max(PSTHs[unitkey][oo]),color='k',linestyle='--')
        ax.vlines(gratingOff,0,max(PSTHs[unitkey][oo]),color='k',linestyle='--')
        ax.set_ylabel('spike count per bin')
        ax.set_xlabel('time in seconds')
        ax.set_title('Channel = '+str(int(unitkey))+' Orientation = '+str(oo))
    plt.tight_layout()
    plt.show()
```

MATLAB

```matlab
f = figure; %1 set(f,'color','w') %2
for cc = 1:numChannels %12
 for oo = 1:numOri %10
  q = subplot(2,1,oo);
  h = plot(timeBase,squeeze(PSTHs(cc,oo,:)),'linewidth',3); %3
  line([gratingOn gratingOn],[min(ylim) max(ylim)], ...
       'color','k','linestyle','--') %4
  line([gratingOff gratingOff],[min(ylim) max(ylim)], ...
       'color','k','linestyle','--') %4
  xlim([min(timeBase), max(timeBase)]) %5
  xlabel('time in seconds') %6
  ylabel('spike count per bin') %6
  title(['Channel = ', num2str(cc), ' - Orientation = ', ...
  num2str(ori(oo)), char(176)]) %7
  set(gca,'Fontsize',14) %8
  q.YLabel.Position(1) = -0.3; %9
 end %10
shg %11
pause %11
end %12
```

To fully understand this code, we have to talk about squeezing. In MATLAB, the function `squeeze` collapses over singleton dimensions in an array. In line %3, we make use of this to plot a particular PSTH over the timebase, particularly the ccth and ooth vector (of 270 entries). Convince yourself that the ccth and ooth PSTH has an extra singleton dimension by typing

```
>> temp = PSTHs(cc,oo,:);
```

then `>> size(temp)`

which should yield

```
ans =     1     1     270
```

for any valid value of cc and oo.
In contrast, `>> size(timeBase)`
yields ans =   1   270
This mismatch in dimensionality will confuse plot and throw an error. We need to remove the extra singleton dimension, which is what `squeeze` is for:

```
>> temp = squeeze(PSTHs(cc,oo,:)); size(temp)
ans =   270     1
```

Now, the dimensions match up and plotting is not a problem. In general, `squeeze` removes all extra singleton dimensions (those of dimensionality one), in this case one extra singleton.

If you did everything right, this code produces figures for all channels. We show data for just one exemplary channel (channel 5) here (Fig. 4.3).

Both types of stimuli produced strong onset transients, but not an appreciable sustained response. Note that the visual aesthetics of the figures this code produces leave a lot to be desired, for instance the axis tick marks are on the inside, the axis labels are too small, and so on. We could fix this now, or, as we will, address these issues later. The reason for that is that this is "just" a data browser. Its explicit purpose is for you to get a feel for the data. It's a lot of effort to make publication-quality figures programmatically and you probably only want to invest that time when making figures that you actually will include in a paper.

Also note that the y-axis has not been normalized—these are raw spike counts per bin. The absolute number of spikes plotted might differ between the two grating conditions, even if the overall shape is quite similar, as in channels 36 or 38.

For now, look at the peristimulus time courses, one channel at a time—and appreciate the tremendous diversity of responses that was captured by the array in cortical area V4.

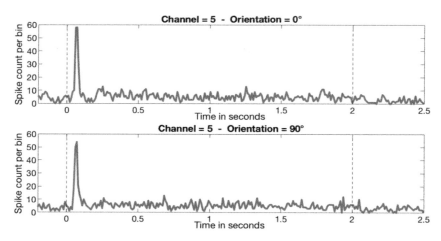

FIGURE 4.3 PSTHs for channel 5. Top panel: Grating orientation = 0 degrees. Bottom panel: Grating orientation = 90 degrees. Hashed vertical lines indicate stimulus on- and offset.

Some of the channels seem to be "dead" altogether—this is unavoidable when inserting an array with a pneumatic hammer—as far as we can tell, channels 21, 31, 34, 56, and 68 are so afflicted. Confirm that this is the case.

Here, we want to show the overall shape of nine classes of neurons existing in this dataset—as far as we can tell. In order to amplify what the shape itself looks like, we don't even plot an axis, you can take the axis off with the command `axis off` in MATLAB and `plt.axis('off')` in Python.

In Fig. 4.4, you can see exemplary time courses of these nine classes—signals on all electrodes exhibit and fall into one of these time courses (outside of those that are dead or very noisy):

a. Transient responses, without appreciable sustained response and no clear offset response.

b. Sustained responses—in addition to the transient, the response is sustained as long as the grating is on.

c. Inverted responses. Neuron seems to *deactivate* upon stimulus onset and go back up to baseline upon cessation of visual stimulation.

d. Ramping behavior—this is a response type we would expect in LIP (Mazurek, Roitman, Ditterich, & Shadlen, 2003; Huk & Shadlen, 2005) and there only when averaging across trials (Latimer, Yates, Meister, Huk, & Pillow, 2015).

e. On- and offset responses, but no sustained response.

f. Offset response only.

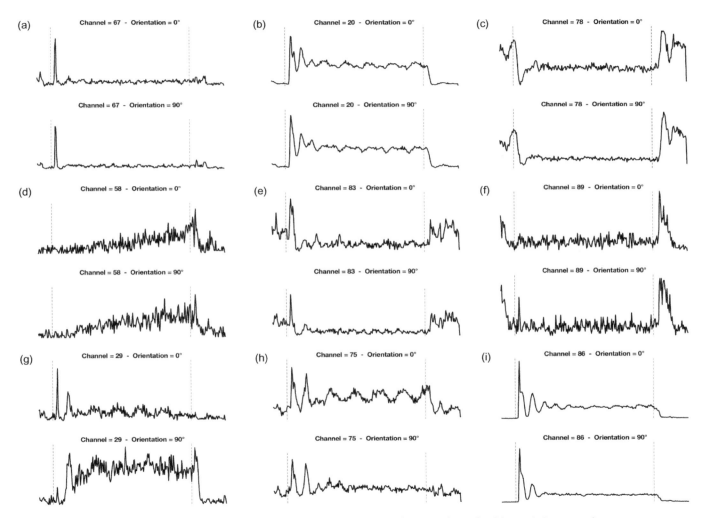

FIGURE 4.4 The bestiarum of response classes in our dataset. Note the dramatic diversity shown by this casuistic approach.

g. Inconsistent. Usually subsumed under "tuning"—the neuron responding to one, but not the other, stimulus or differentially to stimuli in general, but these response profiles seem to be different in kind.
h. Phase locking neurons. These neurons seem to discharge in an oscillatory, sinusoidal fashion, perhaps phase-locked to the stimulus—a moving sinusoidal grating, as a simple cell in V1 would (Mechler & Ringach, 2002).
i. Mixed responses. Neurons like this exhibit combinations of the response classes we described above. For instance this one has some transient response, but also some sustained response and a phasic component.

This example illustrates the value of exploratory data analysis and of looking at the raw data without too much processing. There is also a cautionary tale here. Often, investigators simply average over the stimulus time in the trial to compute the spike rate, then compare that rate across different conditions. It should be obvious simply from looking at these traces that this approach leaves a lot of information on the table—perhaps relevant information.

A second point on this issue is that what allowed us to reveal this diversity is the use of a multielectrode array. Most of the physiological literature relies on recordings from single electrodes where the experimenter decides which neurons to record from and also tailors the stimulus to the response preferences of that neuron. It has been pointed out that these choices might constitute selection bias—experimenters being prone to record from neurons that respond briskly and in a fashion that makes sense, underestimating the neural diversity inherent in the population (Olshausen & Field, 2004, 2005).

To avoid confirmation bias, all data analysis should be exploratory in the beginning. Always look at the raw data first. Be suspicious of processed data. Just like with processed food, it relies on the assumption that everything in the processing chain was done with the right intentions and done right. This might not be true. A highly processed format will hide those problems. Primates are visually guided animals, so always start by looking at the raw data. As raw as possible. If there are problems with the dataset (perhaps due to how it was recorded), they are most likely to be revealed at this stage. You can't afford to skip it. If you don't want to look at your data, listen to them instead if you have to.

So much for the exploratory data analysis of raw (or close to raw) neural signals. Let's now move on to calculating a more abstract quantity, namely the *spike count correlation*.

Before we do that, let's discuss what spike count correlations are and why you should care about them.

To develop this concept, let's first revisit a fundamental property of spike trains, namely that the neural response to any given environmental stimulation is stochastic, showing the same stimulus, e.g., moving gratings like in this experiment, to the same neuron repeatedly can lead to dramatically varying neural responses. We already encountered this property in Chapter 3, Wrangling Spikes Trains. It is not due to spike generation, which is almost perfectly deterministic, precise, and stereotypical (Mainen & Sejnowski, 1995), which suggests that the variability results from the action of other neurons around the one we happen to be recording from, in other words the network that it is embedded in. See Fig. 4.5 for an illustration of this variability.

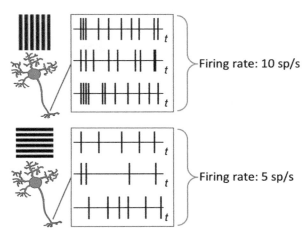

FIGURE 4.5 The variability of spike trains. A given neuron responds differently to different stimuli, e.g., this one discharges at twice the rate in response to a vertical grating as it does in response to a horizontal one, but it does so in a fashion that resembles a poisson process—the spike trains in response to individual presentations (successive horizontal lines) are quite dissimilar from each other.

We have already encountered this basic situation before, in Chapter 3, Wrangling Spikes Trains. Recording from a single neuron with an electrode allows us to quantify properties of the spike train, particularly its mean response, variability, and response latency, which we use to characterize the neuron itself.

Here, the situation is more complex. The data originated from a multielectrode array with 96 channels, allowing to record from many neurons simultaneously. In the interest of clarifying what is going on, we zoomed in on two neighboring recording sites—imagine this situation multiplied by a factor of 24 in both directions (it is a square array) to get the full picture (see Fig. 4.6). In a way, this illustration is more apt than depicting the entire array because almost all of our current methods boil down to characterizing pairwise relationships between two neurons, which does seem to capture much of what is going on in the population (Nirenberg & Victor, 2007). There has been pioneering work to characterize the joint relationship between three neurons, but even adding a single neuron complicates matters tremendously (Ohiorhenuan et al., 2010; Ohiorhenuan & Victor, 2011; Cayco-Gajic, Zylberberg, & Shea-Brown, 2015). Generally speaking, estimating the information content in neural populations, and how much is carried by single neurons versus pairs of neurons versus higher-order arrangements, is tricky (Crumiller, Knight, Yu, & Kaplan, 2011).

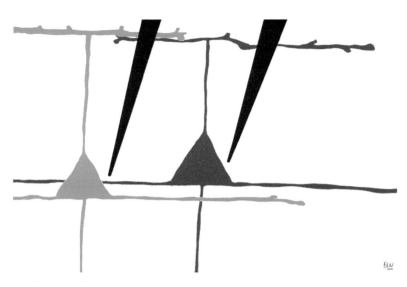

FIGURE 4.6   The colored pyramid shapes with tentacle-like projections are two different neurons, while the black sharp shapes are electrodes that measure the voltage near the neurons.

    This allows us to do more complex analyses. Specifically, we can now quantify the relationship *between* time series or spike counts, in response to the *same* stimulus. This is the spike count correlation, see Fig. 4.7 for an illustration.

    Now, why should you care about this quantity? At first glance, it might seem rather obscure, but its magnitude has profound implications for the question we already raised above—how much information about what the population of neurons is doing is contained in the firing rate of single neurons and the activity of pairs of neurons? This is relevant because, if it is even modestly high, it will put a low ceiling on the information contained in populations of many billions of neurons (Zohary, Shadlen, & Newsome, 1994)—as high spike count correlations imply a tremendous degree of redundancy in neural firing. To get an intuition for this, you could consider what would happen if the average correlation reached 1. In this case, no matter how many neurons in your brain, you could predict all of their activity levels by recording from just a single neuron—because in effect your brain would only contain a single neuron. In contrast, if spike count correlations were zero, there would be no redundancy of information, but predicting the behavior of an animal by recording from a handful of neurons would be hopeless. This is implausible, as the brain harbors a highly

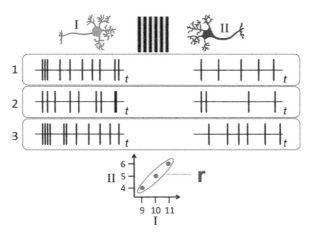

FIGURE 4.7 Spike count correlations. This cartoon depicts the response of two neurons to a vertical grating, recorded simultaneously. The spike trains represent three successive trials. We can then plot the spike counts per trial versus each other. So the *x*-axis represents spike counts for neuron I whereas the *y*-axis represents spike counts for neuron II, for a given trial. Each dot in the plot represents the joint activity for a given trial. The correlation "r" between these two (usually, a Pearson correlation), is the spike count correlation.

connected network of neurons. Empirically, most papers in most neural areas report a spike count correlation of around 0.2–0.3 (Reich, Mechler, & Victor, 2001; Kohn & Smith, 2005; Smith & Kohn, 2008; Gutnisky & Dragoi, 2008), which are thought to reflect either shared bottom-up inputs (Van Kan, Scobey, & Gabor, 1985) or joint top-down influences such as attentional states or behavioral goals of the organism (Nienborg & Cumming, 2009). These theoretical accounts are not mutually exclusive. Some have reported spike count correlations of around zero in primary visual cortex (Ecker et al., 2010), but it has been pointed out that this might be due to unusually low spike rates—spike count correlations are not independent of firing rate (Cohen & Kohn, 2011).

Spike count correlations have dramatic implications for the ability of a neural population to decode stimulus identity. In essence, the perceptual system is in the business of providing the organism with an idea of what the environment is like, so that the motor system can act on it more efficiently. But it is dark in the brain. The skull creates its very own version of Plato's cave—the brain has to figure out what is going on outside of the skull just based on its own activity patterns—i.e., all the information that is available to it.

If spike count correlations are low, determining stimulus identity from neural firing patterns is relatively straightforward (see Fig. 4.8).

As you can see from this figure, neural decoding is straightforward, even for a single trial. The neural responses form clusters because we have already know that there is considerable stimulus response variation from trial to trial. But this is not a problem. If neuron I fires a few spikes and neuron II fires a lot of spikes, it is likely that the stimulus was a vertical grating. Conversely, if neuron I fires a lot of spikes and neuron II fires only a few spikes, it is likely that a horizontal grating was presented. This categorization can easily be implemented by a linear classifier, represented by the blue line in this space, even just on the basis of the activity of two neurons.

But what if the spike counts are correlated?

This situation dramatically complicates the decoding problem, as you can see in Fig. 4.9.

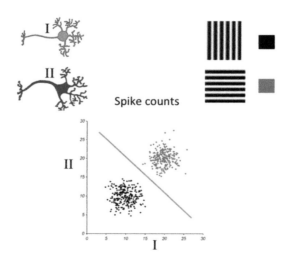

FIGURE 4.8   Decoding without spike count correlations. This cartoon figure depicts the responses (in terms of spike count) of neuron I (plotted on the x-axis) versus neuron II (plotted on the y-axis) to two gratings oriented vertically and horizontally. These gratings are presented repeatedly and each dot represents the response of the two neurons during a given trial—black dots show the response of the neurons to vertical gratings and red dots the response of the neurons to horizontal gratings.

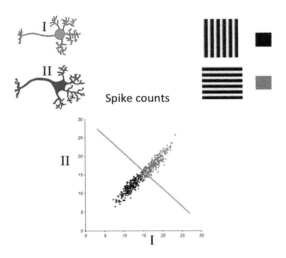

Spike counts

FIGURE 4.9 Decoding with spike count correlations.

Now, decoding stimulus identity is a lot harder and might be limited in principle on a trial-by-trial basis—no matter how many neurons we include in this analysis, there will be mistakes—as they don't add independent pieces of information. Put simply, pooling information from many neurons doesn't help much and decoding accuracy saturates quickly (Zohary et al., 1994). As you can see, the shape of the spike count correlations matters as well—if the clusters were elongated parallel to the decision boundary, these correlations would *aid* in decoding, not limit it (Averbeck, Latham, & Pouget, 2006). We will revisit these issues in Chapter 9, Classification and Clustering, where we will discuss issues of classification in more detail.

First, we need to compute them, which we will do now.

We start by calculating the spike count correlation across all channels—separately for each stimulus type. We use the entire signal on each electrode, with an understanding that this analysis includes multiunit samples where multiple neurons potentially contribute to the signal on each electrode. But before we do so, we need to eliminate the five "dead channels," which are channels [21, 31, 34, 56, 68].

Pseudocode

---

0. Copy the variable PSTH into livePSTHs - note the 0-indexing here
1. The vector of dead channels without signals
2. Removing them from the PSTHs matrix
3. Calculating the remaining number of channels

Python	MATLAB

```
livePSTHs = PSTHs #0
deadchannels = [21, 31, 34, 56, 68] #1
[livePSTHs.pop(dead, None) for dead in deadchannels ] #2
numChanLeft = len(livePSTHs) #3
```

```
livePSTHs = PSTHs; %0
Nothing = [21, 31, 34, 56, 68]; %1
livePSTHs(Nothing,:,:) = []; %2
numChanLeft = size(livePSTHs,1); %3
```

The reason we first copy the PSTHs matrix and assign it to the matrix livePSTHs is that once we delete the five dead channels, we'll "lose our spot"—if we want to look at some individual channels afterwards, using the old numbering, this will no longer work. It's generally a good idea, unless the variable is simply too big, to keep copies of data when deleting some of it, particularly if the deletion changes the structure of the data file and thus affects all of the data in it, even the good data.

Now, on to calculating the spike count correlations:

Pseudocode

---

1. Allocate the matrix rSC that will contain the spike count correlations. Initialize it with zeroes.
2. Correlate the signal on a given channel with another one, for a given orientation
3. Assign the resulting correlation to the right place in the rSC matrix
4. Loop over all orientations
5. Loop over all column channels
6. Loop over all row channels
7. Calculating the overall spike count correlation in this dataset

---

Python	MATLAB

```
rSC = np.zeros(shape=(len(livePSTHs),len(livePS
THs), len(ori))) #1
for rowind,rowkey in enumerate(livePSTHs.keys()):
#6
    for colind,colkey in enumerate(livePSTHs.
keys()): #5
        for oind,oo in enumerate(ori): #4
            rSC[rowind,colind,oind],dummy =sc.
pearsonr(livePSTHs[rowkey][oo], livePSTHs[colkey]
[oo]) #3
globalMean=np.mean(rSC) #7
```

```
rSC = zeros(numChanLeft,numChanLeft,numOri); %1
for rows = 1:numChanLeft %6
 for columns = 1:numChanLeft %5
  for oo = 1:numOri %4
   temp ...
corrcoef(livePSTHs(rows,oo,:),livePSTHs(columns,oo,:)); %2
                rSC(rows,columns,oo) = temp(1,2); %3
  end %4
 end %5
end %6
globalMean = mean(mean(mean(rSC))) %7
```

This code returns a `globalMean` (the overall *spike count correlation*) of 0.153 between the signals on all channels. Not bad, given that we did no other preprocessing, thresholding, or other selection of signals. Some of the channels are quite noisy, lowering the overall correlation.

Is there a fine structure to this overall result? To find out, it is perhaps best to simply visualize the pairwise correlation between all channels at once, something we are already well positioned to do:

Pseudocode

---

0. Import package
1. Open a new figure
2. Open a new subplot, one for each type of orientation
3. Image all channels, per orientation
4. Make sure that the axis is square to avoid aspect ratio distortions
5. Set colormap to "jet"
6. Add a colorbar to get a sense for the magnitude of correlations
7. Add a suitable title
8. Make font big enough to read ticks and title
9. Loop over orientations

Python | MATLAB

```python
rSC = np.zeros(shape=(len(livePSTHs),len(livePSTHs),
len(ori)))
for rowind,rowkey in enumerate(livePSTHs.keys()):
    for colind,colkey in enumerate(livePSTHs.
keys()):
        for oind,oo in enumerate(ori):
            rSC[rowind,colind,oind],dummy =sc.
pearsonr(livePSTHs[rowkey][oo], livePSTHs[colkey]
[oo])
globalMean=np.mean(rSC)
```

```matlab
figure %1
for oo = 1:numOri %9
    subplot(1,numOri,oo) %2
    imagesc(rSC(:,:,oo)) %3
axis square %4
colormap(jet) %5
colorbar %6
title(['Orientation: ', num2str(ori(oo)), char(176)]); %7
set(gca,'fontsize',16) %8
end %9
```

Voila (Fig. 4.10):

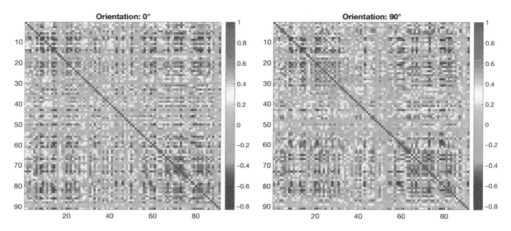

FIGURE 4.10   Pairwise spike count correlations between all live channels. Red areas indicate units whose pair-wise correlations are near 1, meaning very correlated. Green areas indicate units whose correlation values are near 0, meaning they are not correlated. Blue areas indicated units within correlations near –1, or *inversely* correlated. As you can see, they are mostly positive, but with some green and blue valleys. The latter probably corresponds to channels with "inverting" responses that are inversely correlated with those that exhibit "sustained" responses. A spot-check suggests that this might well be the case, e.g, the correlation between channels 20 and 78 is a whopping –0.7805.

What about the idea that some noisy channels drop the magnitude of the overall correlation? By eyeballing, we note that some channels don't exhibit much of a response to the stimulus or only have a few spikes, making it hard to tell what is going on, namely these channels:

Noisy = [7, 16, 32, 37, 39, 41, 43, 45, 47, 48, 51, 52, 54, 94, 95].

If we remove them from the analysis, there are 76 channels remaining and the mean spike count correlation goes up to 0.185. There you go, much less green, witness Fig. 4.11.

We can use some additional information about the experiment to arrive at a better estimate of the spike count correlation. The grating was on for 2 s, or 200 bins. We wouldn't expect there to be much synchronization of activity (correlated or anticorrelated) before grating onset and after grating offset.

We can take this into account by only correlating the activity from grating onset to grating offset by adding beginning and end time bins to the MATLAB code:

```
temp = corrcoef(livePSTHs(rows,oo,11:211),livePSTHs(columns,oo,11:211));
```

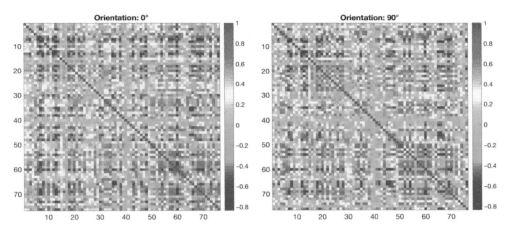

FIGURE 4.11   Pairwise spike count correlations with noisy channels removed.

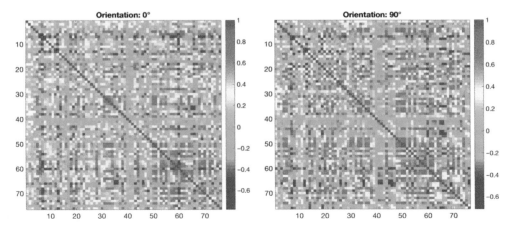

FIGURE 4.12   Correlations calculated during stimulus presentation.

This choice of bins is reasonable and brings the overall spike count correlation up to 0.2. It is particularly high in response to the 90 degree grating (0.22) versus the 0 degree grating (0.18), as you can see in Fig. 4.12.

Finally, recall that we used information from multiunit samples (all spikes from a given electrode). We could use the sort codes in the original data to further refine the analysis—by unit, not by channel.

Doing so, we realize that most channels have a single unit, so we don't expect too much improvement, but some channels (e.g., channel 20) have up to four units, so perhaps we'll see something. A by-unit analysis has 106 units, but we have our noise channels back, so the mean spike count correlation is back down to 0.15 as you can see in Fig. 4.13.

Eliminating units that fired very low rates and/or showed no discernible response to the stimulus, we can get back a somewhat more defined pattern, with a global mean spike count correlation of 0.189 (see Fig. 4.14).

Getting here required such extensive code surgery—we won't rewrite the code here, in order to save space. You can download it from the companion website.

One of the reasons that looking at individual units didn't improve the overall spike count correlation could be that PSTHs are noisy for some neurons that don't fire much. This noise will reduce correlations and make overall determination of what is going on hard. This is why some researchers smooth their spike trains by convolution before doing any further analysis, in particular when firing rates are low and bin sizes are narrow.

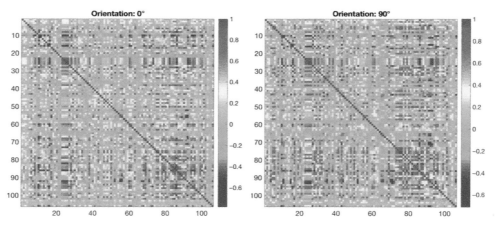

FIGURE 4.13   Expanding correlation analysis from multiunit to single unit.

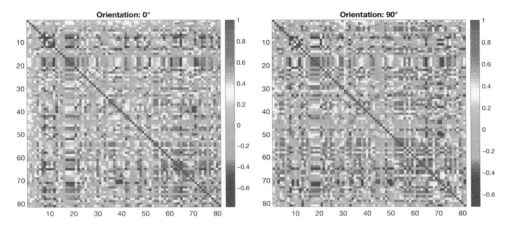

FIGURE 4.14 Single unit correlation analysis with noisy channels removed.

Unit 6 (in the new by-unit analysis) is a good example of this. It is clear enough what is going on, and the onset transient is strong. However, in 1150 trials, less than 60 spikes fell even into the peak bin, for an instantaneous firing rate of 5 spikes per second, which is not awfully high. Most of the wiggles in the "sustained" portion of the response, while the grating is still on, are probably not real, but dominated by noise, which will throw off the correlational analysis (most of the response is unreliable). Smoothing the signal will mitigate this problem, and is a popular method for improving correlation analyses (Fig. 4.15).

Here, we *convolve* the signal with a *kernel* specified as 5 bin-width long and uniform, meaning that the kernel is an array of five 1's. The variable `kernel`, when convolved with a PSTH, acts as a sort of smoothing filter. We perform this *convolution* using the POM functions `np.convolve` and `conv` by adding these lines to the code, then plotting `convT` and `convS` in the data browser:

Pseudocode

---

1. Specify kernel of length 5
2. Calculate the convolution of the PSTH with the kernel
3. Calculate the convolution of the time vector with the kernel

---

Python	MATLAB

```
kernel = np.ones(5)
convS = np.convolve(PSTHs[unitkey][oo],kernel)
convT = np.convolve(timeBase,kernel)
```

```
kernel = ones(5,1);
convS = conv(squeeze(PSTHs(oo,cc,:)),...
kernel,'valid')./sum(kernel);
convT = conv(timeBase, kernel,'valid')./sum(kernel);
```

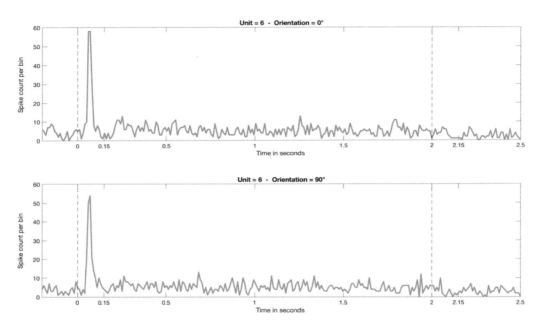

FIGURE 4.15   Unit 6 before convolution.

Considerably smoother. Note that the firing rate in the peak bin has decreased considerably. This is because it has been blended with the values of neighboring bins, which we know to be much less responsive. Most convolution of spike trains is done on bin-widths of 1 ms with narrow kernels that are of Gaussian shape. But we just want to illustrate the concept here (Fig. 4.16).

This would make a good flag to set at the end of the initialization block of the code—set a variable that determines whether convolution should be done or not to 0 or 1, then only execute the convolution if the flag is on, e.g., by adding the line doConv = 1; perhaps also add parameter to make the bin width flexible.

As a conceptual understanding of convolution is useful in Chapter 5, Analog Signals, as well (and convolution is a basic tool in the toolkit of every neuroscientist) it behooves us to introduce it here briefly.

We recognize that we are just about to finish a heavy chapter. If you want, you can either treat this as a grand finale, or, if exhaustion has set in, start over here later and treat this as the introduction to Chapter 5, Analog Signals.

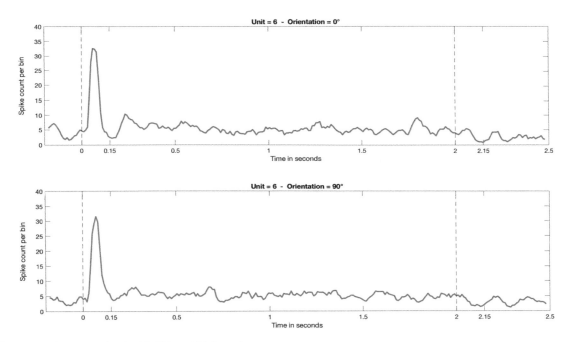

FIGURE 4.16 Unit 6 PSTH convolved with 5 bin-width kernel.

Most attempts to explain convolution are themselves rather convoluted—we'll try to avoid that here.

For our purposes, convolution boils down to a moving average—the multiplication of each value in a time series with a corresponding value in a "kernel" that slides across the time series, then summing up all of these products to yield a new value (often divided by the sum of all numbers in the kernel to get an average, instead of a sum; see Fig. 4.17). In this figure, we slide kernels of equal weights along a time series.

You can see most of what is relevant about convolution (for our purposes), in this image. Specifically:

1. Convolution amounts to a smoothing of the time course.
2. Convolution begins as the first element of the kernel falls on the first element of the time series. Values that don't align are "zero-padded," i.e., the kernel value at that point is multiplied by zero.

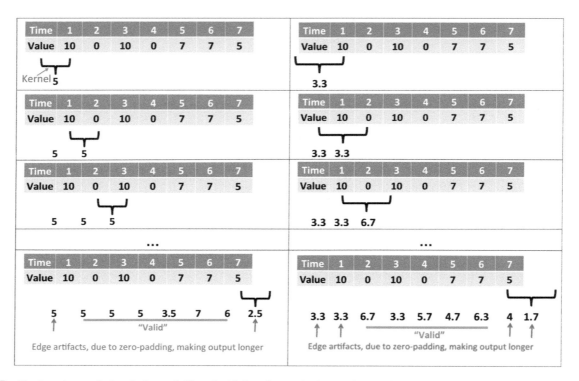

FIGURE 4.17   The joy of convolution. Left panel: Kernel with length two. Right panel: Kernel with length three. Rows represent successive moves of the kernel. The kernel is represented as curly braces rotated by 90 degrees. Kernel values are [1 1] on the left and [1 1 1] on the right.

3. If you zero-pad, the vector that results from convolution will be n+ (k–1) elements long, where n is the number of elements of the time series and k is the number of elements of the kernel.
4. If you take only "valid" results, i.e., positions in which the entire kernel falls on the time series and that don't involve zero-padding, the time series is shorter, n–(k–1) elements long.
5. Even kernel lengths will amount to a phase shift of the resulting vector, whereas an odd-numbered kernel length produces results from the middle of the kernel and doesn't introduce a phase shift.

6. Generally speaking, longer kernels smooth the time course more strongly, but this depends on the structure of the data itself. If the periodicity of the fluctuations in the time series aligns with the length kernel (as in the first part of the time series here), a shorter kernel can produce a smoother output.

Convolution works by shifting the kernel along the time series, then multiplying corresponding elements and summing them up. For a deeper appreciation, you can download an animation that shows the sliding kernel in action from the companion website. The method is agnostic about which of the two vectors is in motion. You can conceptualize this either as the time series moving against a static kernel or a moving kernel sliding over a static time series—it makes no difference to the results. Historically, it is known to be hard to tell which of two systems are in relative motion, if situated in one of the two systems.

Because we used simple uniform kernels with equal weights, this operation corresponds to just a moving average. If the weights of the kernel had been different, this would have corresponded to a moving weighted average (technically speaking, a shifting dot product between time series and kernel). A popular choice for the shape of the kernel in neuroscience and engineering is a Gaussian to minimize edge artifacts.

If we run or rerun our entire analysis on the convolved PSTHs with a kernel width of 30ms, we achieve the highest mean spike count correlation yet: 0.21. Note that this value deemphasizes the changes, as the blue regions corresponding to negative correlations have also become more pronounced and the correlation plot looks striking. If we had kept track of the absolute value of correlations, every step we took would have increased them, on average (Fig. 4.18).

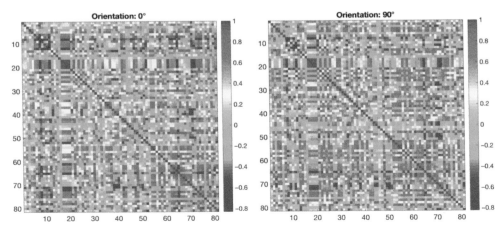

FIGURE 4.18 Pairwise spike count correlations after convolving PSTHs with kernel.

In terms of our 0 to 1 approach, we now have reached 1 as far as calculating spike count correlations are concerned. Going from 1 to 100 (and beyond) could take any number of shapes, such as looking at spike count correlations in relation to other things, such as spatial distance (Smith & Kohn, 2008) or tuning similarity between neurons (Kohn & Smith, 2005), or the relationship between spike count correlations and the LFP (Snyder et al., 2015). Speaking of the LFP: Let's move on to Chapter 5, Analog Signals, where we will go from 0 to 1 in signal processing of LFPs.

⌘. *Pensee on whether nested loops are a necessary evil:* We think that nested loops are neither necessary, nor evil. They got their bad reputation because they are slow and inefficient. Professional programmers abhor them because—as they are almost always unnecessary—they are a sure sign of a rookie programmer who doesn't know what they are doing. However, it is important to keep perspective. They might well be slow and inefficient, but they get the job done. POM reads, and executes, the code line by line, which is what makes nested loops so slow, but at the same time extremely comprehensible for people reading through the same code. This is why we are using it at this point, when introducing these concepts in a primer. We'll later show how the very same code can be vectorized, which radically speeds up execution time of the code, but makes it hard to comprehend and maintain. In an age of ever-increasing processor speeds, we think nested loops are justifiable. Just don't overdo it. If you have a six-deep nested loop, e.g., looping through all experiments, all conditions, all animals, all trials, all electrodes, and all spikes at once, you should anticipate some wait time if the dataset is sufficiently large—you might well think you are in limbo, with time essentially being at a standstill, if you go too deep down into the hierarchy of loops.

CHAPTER

5

Analog Signals

In this chapter, we will cover basic signal processing techniques by analyzing neural data in the form of analog signals. Analog signals are commonly used in neuroscience—the electroencephalogram (EEG) records large electric potentials from the scalp, the magnetoencephalogram (MEG) measures magnetic fields from sources within the brain, the local field potential (LFP) is obtained through recording the activity of many neurons; all are analog neural signals.

Looking at spike trains (which we have done in the last two chapters) amounts to looking at a signal that is fundamentally binary and discrete—a neuron either fires, or it does not fire. In other words, we have been looking at inherently *digital* signals. But the underlying voltage trace from which we extracted the spikes was an analog signal, it was a time course of voltage values. Extracting spikes from this analog signal corresponds to feature extraction—spikes are such salient features (high frequency and high amplitude)—that one would be hard pressed not to note the presence of spikes in a voltage trace. In contrast, when working with analog signals, one is typically working with the entire signal, not with detected features. When working with the entire signal, it is helpful to have some basic knowledge about signal processing, which we hope to develop in this chapter.

A starting point in signal processing is the notion that some signals are best represented in the time domain but aspects of others are better represented in the *frequency domain*. The frequency domain (or *frequency space*) is often rather counterintuitive to the beginner, as we all live our lives in the time domain and our data are usually returned by our data recording apparatus in the form of a time series (in the time domain) as well. So what is frequency space and why is a frequency space representation of data useful?

Simply put, a frequency space representation emphasizes periodic or repeating components of a signal. This might sound somewhat abstract, so it is perhaps best to jump right in, with a coded example. We start by plotting *sine waves*.

A sine wave is a cyclically repeating signal that can be completely described by just three parameters: (1) its frequency (how many times the cycle repeats per second), (2) its amplitude—we didn't specify this here, by default the sine wave

Neural Data Science.
DOI: http://dx.doi.org/10.1016/B978-0-12-804043-0.00005-2

goes from −1 to 1 if we multiplied the signal time course by a scaling factor, we could make it go from −0.1 to 0.1 or from −100 to 100 or any other range we would like, (3) its phase, which is the point in the cycle at which we start drawing the sine wave. We didn't specify that here either and by default, the sine wave starts at y-position 0 at time 0. The cosine function is simply a sine wave that is phase-shifted by 90 degrees—it starts to draw at y-position 1 at time 0 (see Fig. 5.1).

Pseudocode

---

1. Open figure
2. Define the sampling frequency, here 1000 Hz
3. Duration of our time series, in seconds. Here: 1 second
4. Define a time base t, in steps of sampling frequency
5. Define signal frequency, here 2 Hz
6. Represent a sine wave
7. Represent a cosine wave
8. Open a subplot with two horizontal panels
9. Plot the sine wave over the time base as a black, thick line
10. Open the a subplot in the lower panel
11. Plot the cosine wave over the time base as a black, thick line

---

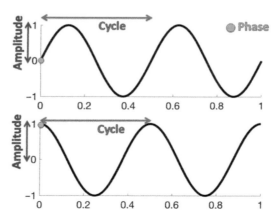

FIGURE 5.1   Sine (top panel) and cosine (bottom panel) function. We added arrows to point out concepts we referenced in the text above, such as amplitude, cycle, frequency, and phase (these are not created in code). The x-axis represents time, so both functions have a frequency of 2 cycles per second or 2 Hz.

Python	MATLAB
| ```
import matplotlib.pyplot as plt
import numpy as np
fig=plt.figure() #1
fs = 1000; #2
dur = 1; #3
t = np.linspace(0,dur,fs*dur); #4
freq = 2; #5
sinW = np.sin(2*np.pi*freq*t); #6
cosW = np.cos(2*np.pi*freq*t); #7
ax=plt.subplot(2,1,1) #8
ax.plot(t,sinW,c='k',lw=4) #9
ax=plt.subplot(2,1,2) #10
ax.plot(t,cosW,c='k',lw=4) #11
``` | ```
figure %1
fs = 1000; %2
dur = 1; %3
t = 0:1/fs:dur; %4
freq = 2; %5
sinW = sin(2*pi*freq*t); %6
cosW = cos(2*pi*freq*t); %7
subplot(2,1,1) %8
plot(t,sinW,'color','k','linewidth',4) %9
subplot(2,1,2) %10
plot(t,cosW,'color','k','linewidth',4) %11
``` |

Now that we know what sine waves are, we can build up a signal that is a bit more complex.

Pseudocode
1. Open a new figure
2. Duration of our time series, in seconds. Here: 4 seconds
3. Define a time base t, in steps of sampling frequency
4. Define signal frequency, here 10 Hz
5. Create the time course of the signal, here a sine wave
6. Open a subplot with three horizontal panels
7. Plot the signal over the time base as a black, thick(ish) line

Python	MATLAB
| ```
fig = plt.figure() #1
dur = 4 #2
t = np.linspace(0,dur,fs*dur) #3
freq = 10 #4
signal = np.sin(2*np.pi*freq*t) #5
ax = plt.subplot(3,1,1) #6
ax.plot(t,signal,c='k',lw=2) #7
``` | ```
figure %1
dur = 4; %2
t = 0:1/fs:dur; %3
freq = 10; %4
signal = sin(2*pi*freq*t); %5
subplot(3,1,1) %6
plot(t,signal,'color','k','linewidth',2) %7
``` |

If you execute this code, you can count the number of cycles—either by counting the number of peaks or valleys, see that the number is 40, divide it by the duration of the signal (4 seconds) and conclude that the signal was a sine wave with a frequency of 10 Hz. Which is true. So far, so good, see top panel of Fig. 5.2.

But signals with a single cyclical component are rare, particularly in neuroscience. Let's see what happens if we add a second sine wave to the first one. We haven't done anything fancy, we didn't change phases or amplitudes (which will also most definitely be variable in real neuroscience applications), but can you still discern the two component signals, even in this simple case?

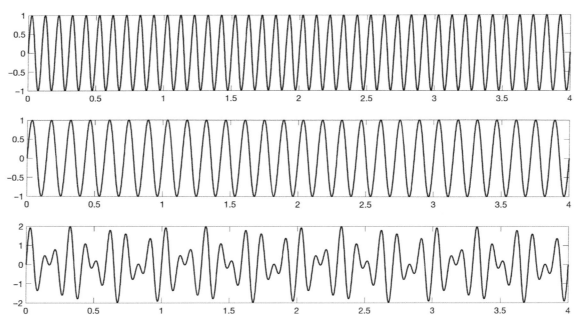

FIGURE 5.2   Adding two sine waves. Top panel: Sine wave with frequency 10 Hz. Middle panel: Sine wave with frequency 7 Hz. Bottom panel: The two sine waves, added; x-axes: time in seconds; y-axes: amplitude.

Pseudocode

---

1. Define signal frequency of second sine wave, here 7 Hz
2. Create the time course of the 2nd signal, again a sine wave
3. Open the 2nd horizontal panel in the subplot
4. Plot the 2nd signal over the time base as a black, thick(ish) line
5. Add the two signals
6. Open the bottom horizontal panel in the subplot
7. Plot the joint signal over the time base as a black, thick(ish) line

---

Python	MATLAB

```
freq2 = 7; #1                              freq2 = 7; %1
signal2 = np.sin(2*np.pi*freq2*t); #2      signal2 = sin(2*pi*freq2*t); %2
ax=plt.subplot(3,1,2) #3                   subplot(3,1,2) %3
ax.plot(t,signal2,c='k',lw=2) #4           plot(t,signal2,'color','k','linewidth',2) %4
jointSignal = signal+signal2; #5           jointSignal = signal+signal2; %5
ax=plt.subplot(3,1,3) #6                   subplot(3,1,3) %6
ax.plot(t,jointSignal,c='k',lw=2) #7       plot(t,jointSignal,'color','k','linewidth',2) %7
```

Unless you are an expert in signal processing, we wager that you may not be able to. Call it Pascal's wager (reloaded). And this is a signal with just two pure sine waves, with the same amplitude and no relative phase shifts.

In addition, pure sine waves are a platonic concept: they only live in the realm of ideas. The real world, in which we all dwell, is messy and dirty. The real world is corrupted by *noise*. Always. Whereas a black hole might or might not have hairs (Gürlebeck, 2015), anything in neuroscience most definitely has hairs and lots of it, so we better get used to handling this abundant hairiness. This starts now, right away. Let's add just a sprinkle of noise to the signal:

Pseudocode

---

1. Determine the number of points in the time base
2. Create a vector of random numbers drawn from a normal distribution, with the same length as the signal (and time base)
3. Add the noise to the signal
4. Open a new figure
5. Open a new subplot
6. Plot the joint signal from 5.2 again
7. Switch to middle panel of subplot
8. Plot the signal with the noise over the entire time base

(Continued)

Python	MATLAB
`n = len(t) #1`	`n = length(t); %1`
`noise = np.random.randn(n) #2`	`noise = randn(1,n); %2`
`signalAndNoise = jointSignal + noise #3`	`signalAndNoise = jointSignal + noise; %3`
`fig = plt.figure() #4`	`figure %4`
`ax = plt.subplot(3,1,1) #4`	`subplot(3,1,1) %5`
`ax.plot(t,jointSignal,c='k',lw=2) #6`	`plot(t,jointSignal,'color','k','linewidth',2) %6`
`ax = plt.subplot(3,1,2) #7`	`subplot(3,1,2) %7`
`ax.plot(t,signalAndNoise,c='k',lw=2) #8`	`plot(t,signalAndNoise,'color','k','linewidth',2) %8`

This is starting to look a lot like a sleep EEG. At this point, even seasoned veterans of signal processing will be hard pressed to resolve the component signals in this time course. That's ok—it is extremely hard to recover the cyclical components of even simple signals in a moderate amount of noise by looking in the time domain. In perception, perspective matters. That's where a frequency domain representation comes in. It highlights cyclical components in a signal, making it much easier to ascertain which of them are present—or absent.

To see the power of looking at the frequency domain (literally), we'll look at this signal in frequency space right now. You may not fully understand the code that we are writing at this time, but that is ok—we will use this gap to motivate how to close it in the remaining text of this chapter. Perhaps this code will elicit a burning desire to understand it in you, the reader. If so, we have succeeded. It should be clear what exactly is going on by the end of this chapter.

Pseudocode	
	1. Determine the Nyquist frequency
	2. Do the fast Fourier transform of the combined signal, normalized by time
	3. Create a frequency base for plotting, in analogy to the time base that goes from zero to the Nyquist frequency and is a long as there are valid points in the fft
	4. Switch subplot to the last panel, call it powerPlot
	5. Determine what half the signal in frequency space is
	6. Take the complex conjugate of that
	7. Calculate power
	8. Plot power over the frequency base
	9. Pick appropriate x and y limits of the plot—we could in principle look at everything from 0 to 500, but the two values we care about would be bunched too close together
	10. Add tickmarks with a higher resolution
	11. Add axes labels

(Continued)

| Python | ```
nyquist = fs/2 #1
fSpaceSignal = np.fft.fft(signalAndNoise)/len(t) #2
fBase = np.linspace(0,nyquist,np.floor(len(signalAndNoise)/2)+1) #3
powerPlot = plt.subplot(3,1,3) #4
halfTheSignal = fSpaceSignal[:len(fBase)] #5
complexConjugate = np.conj(halfTheSignal)#6
powe = halfTheSignal*complexConjugate#7
powerPlot.plot(fBase,powe, c='k',lw=2) #8
powerPlot.set_xlim([0, 20]); #9
powerPlot.set_xticks(range(20));#10
powerPlot.set_xlabel('Frequency (in Hz)') #11
powerPlot.set_ylabel('Power')#
``` |
|---|---|
| MATLAB | ```
nyquist = fs/2; %1
fSpaceSignal = fft(signalAndNoise)/(length(t)/2); %2
fBase = linspace(0,nyquist,floor(length(signalAndNoise)/2)+1); %3
powerPlot = subplot(3,1,3) %4
halfTheSignal = fSpaceSignal(1:length(fBase)); %5
complexConjugate = conj(halfTheSignal); %6
pow = halfTheSignal.*complexConjugate; %7
h = plot(fBase,pow, 'color','k','linewidth',2) %8
xlim([0 20]); ylim([0 1]); %9
powerPlot.XTick = 0:20; %10
xlabel('Frequency in Hz'); ylabel('Power'); %11
``` |

The figure produced by all this code should look something like Fig. 5.3—although not precisely because we add some random noise that impacts how the last two panels look.

Before we move on, we want to note several things about the code and the plot.

The heavy lifting in this function is done by the Fourier transform, specifically a fast version of it, developed by Tukey (Cooley & Tukey, 1965). It is the Fourier transform that transports us (or rather, the signal) into frequency space. Frequency space is a place of magic and wonder—the Fourier transform of a time series usually involves complex numbers with imaginary parts. That is why we plot the Fourier transform itself by typing plot(fSpaceSignal), it will yield a funky plot (try it). That's because the complex numbers represent both *magnitude* and *phase* at once, but we are only interested in amplitude or power at this point, which is why we plotted only the absolute values above. Also, if you type plot(abs(fSpaceSignal)), you will get a plot of all the magnitudes, but note that it is mirror-symmetric in the middle—it repeats after half the sampling rate, which is why we plotted things only up to half the sampling rate above (Fig. 5.4).

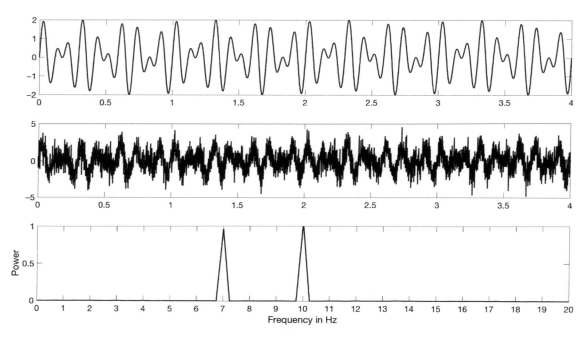

FIGURE 5.3 Adding noise to the sine waves makes it hard to discern what is going on with the signal in the time domain but not in the frequency domain. Axis labels are time vs. amplitude in the first two plots and frequency vs. power on the last one. Note that the power is not exactly 1 for each component, as noise doesn't just add power to where there is none (frequencies other than 7 and 10), it also removes power from where there is signal (7 and 10).

As you can see, power matters. For cyclical phenomena, always have the will to look at power.

If you go beyond half the sampling rate (also called the *Nyquist frequency* in honor of Harry Nyquist), you will encounter a phenomenon called *aliasing*. Put simply, the shape of signals over time can only be represented properly if one is sampling the signal more than twice as fast as the fastest frequency component in the signal itself. Otherwise, the shape of the sampled signal will be distorted. This might appear somewhat abstract, but because it is such a fundamental principle of signal processing, it is worth trying to really understand it—much of what follows will be dependent on it, and we'll try to do so with an example. Again, let's first write the code, then look at the figure, then try to understand what is going on.

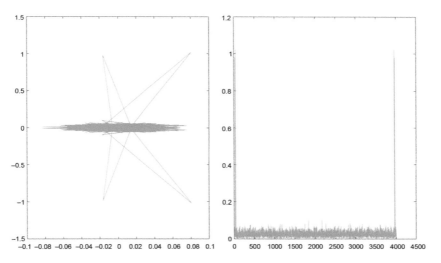

FIGURE 5.4 If you see this, you are doing it wrong. Left panel: `plot(fSpaceSignal)`. Right panel: `plot(abs(fSpaceSignal))`.

Pseudocode

1. Define the frequency of reality, here 1000 Hz
2. Define a time base with the same resolution
3. Define the signal frequency, here 10 Hz
4. Pick 9 strategically chosen sampling frequencies to show something interesting
5. Define the signal as a sine wave at the signal frequency over time
6. Open a figure (for Fig. 5.5)
7. Open a subplot
8. Plot the signal over time
9. Turn the hold on so that we can plot in the same subplot again without overwriting the old one
10. Calculate the sampled signal for a given sampling rate
11. Plot the sampled signal over the same time base in red
12. Add black circles as markers, with yellow edges
13. Make them a bit bigger so one can see them
14. Add a title, namely the corresponding sampling frequency, assign to title handle hT
15. Make title font larger, set it to 26
16. Loop over all sampling frequencies

(Continued)

Python	
| | ```
realSampling = 1000 #1
t = np.linspace(0,1,realSampling) #2
signalFrequency = 10 #3
samplingRate = [3, 4, 8, 10, 11, 12, 20, 41, 200] #4
signal = np.sin(2*np.pi*signalFrequency*t)#5
fig=plt.figure() #6
for ii,sampleRate in enumerate(samplingRate): #15
    ax = plt.subplot(3,3,ii+1)#7
    ax.plot(t,signal)#8
    sampledSignal = np.rint(np.linspace(0, len(t)-1, sampleRate)).astype(int)#10
    q = ax.plot(t[sampledSignal],signal[sampledSignal],c='r',marker='o',mfc='k',mec='y',
markersize=6); #11
    plt.title('fs = '+str(sampleRate)) #14
``` |
| MATLAB | |
| | ```
realSampling = 1000; %1
t = 0:1/realSampling:1; %2
signalFrequency = 10; %3
samplingRate = [3 4 8 10 11 12 20 41 200]; %4
signal = sin(2*pi*signalFrequency*t); %5
figure %6
for ii = 1:length(samplingRate) %16
subplot(3,3,ii)%7
plot(t,signal)%8
hold on %9
sampledSignal = round(linspace(1, length(t), samplingRate(ii))); %10
q = plot(t(sampledSignal),signal(sampledSignal),'color',' r'); %11
q.Marker = 'o'; q.MarkerFaceColor = 'k', q.MarkerEdgeColor = 'y'; %12 q.MarkerSize = 6; %13
hT = title(['fs = ',num2str(samplingRate(ii))]); %14
hT.FontSize = 26; %15
end %16
``` |

This code yields Fig. 5.5.

The top row in Fig. 5.5. represents cases of serious undersampling. If we sample the signal at 3 Hz, there seems to be no signal, whereas sampling at 4 and 8 Hz seems to yield one and two cycles, respectively. When sampling around the signal frequency, interesting things happen, as seen in the middle row, either missing the signal completely or yielding a single sinusoidal cycle of differing frequencies. The bottom row shows sampling at or beyond the Nyquist frequency. Once one is sampling at the Nyquist frequency (here, 20 Hz), the number of cycles in the original signal will be

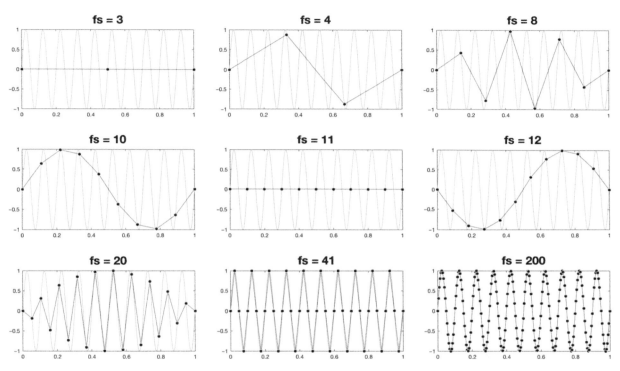

FIGURE 5.5 Aliasing. The blue trace in each subplot is the same and represents the real signal. The red traces correspond to the sampled signal, sampled at the frequency indicated in the title of the book. Black dots represent the positions at which the real signal was sampled.

recovered, but not their amplitude. To get a good estimate of the amplitude as well (try a sampling rate of 23 Hz to see some interesting examples of amplitude modulation), one has to sample at a minimum of twice the Nyquist frequency, or four times the signal frequency. This makes sense because at that frequency, we now have four points (or measurements) per cycle in the original signal—if properly placed, this will allow us to represent everything that matters about the periodic signal. Going much beyond that yields diminishing returns—sampling the signal at 20 times the signal frequency yields only incremental returns over doing it at four times the signal frequency, as seen in the lower right, but it looks much smoother.

What causes a lot of the confusion about sampling is that we have to distinguish three frequencies here. The first frequency we need to consider is that of reality itself. In our toy example, reality is updated a thousand times a second, which is sufficient for our purposes. The reality we live in is (as far as we can tell) updates 1.85×10^{43} times per second (Planck, 1900), or 1 over Planck time (t_p, the time it takes a photon to travel one Planck length, the smallest meaningful unit of time we [science] have been able to discern so far). The second frequency of interest is the periodic frequency of the signal. Here, our signal repeats 10 times per second. You can convince yourself that there are 10 cycles in the 1 second of time we plot. The last question is how often we take a measurement of our signal. If we take two measurements per second, our sampling frequency is 2 Hz. If we take 300 measurements in the same time period, our sampling frequency is 300 Hz. In physiology, it is not unusual for rigs to sample the signal tens of thousands of times per second (about 40 kHz) taking a measurement (in electrophysiology usually of the voltage at the electrode tip) every 25 μs, which will allow for a high fidelity representation of neural signals - as we already discussed in this book, spikes occur on the timescale of about 1 ms, so if you want to resolve the shape of the waveform, e.g. for purposes of spike sorting, you will want to sample the signal several times per ms.

It is this sampled signal that comes out of the *rig* as data and forms the basis of all further analysis, so it is critical to sample the signal at high enough of a rate to fully represent all aspects (frequency and amplitude) of the physiological signals of interest. For reasons outlined above, four times the highest signal frequency or above will do nicely.

Remember this the next time you get into a political argument with someone—you are both commenting on the same reality, but are probably sampling it differently and undersampling at that. Perhaps try to sample at a higher frequency?

So much for aliasing—back to the Fourier transform, the algorithm that does all the work.

The Fourier transform was developed by Joseph Fourier. His use case was to analyze a pressing scientific problem of his time, the heat distribution on a cannon muzzle (Fourier, 1878).

The point of the Fourier transform is that any signal in the time domain can be decomposed into a set of repeating signals with suitable frequencies, amplitudes, and phases (and vice versa—the inverse transform can transport any signal in the frequency domain back into the time domain).

Here is the equation:

$$\Im\{x(t)\} = X(f) = \int_{-\infty}^{\infty} x(t)e^{-i2\pi ft}\,dt$$

(5.1)

EQUATION 5.1   The Fourier Transform.

At first glance, this equation can look a bit scary and intimidating to the untrained eye. John von Neumann observed that "in mathematics, you don't understand things—you just get used to them" (Zukav, 2012). This seems like a copout and we could not disagree more strongly. Of course you could simply trust the mathematicians that it works and keep using it until you get used to it, but we think it is much better to understand it, which is actually possible. To attempt this, let's first decompose it, explain the meaning of each component, then visualize what is going on. On the far left is a fraktur "F," which represents the Fourier transform itself. What is the Fourier transform transforming? A function x. Function x is a function of t (time). This function could represent a signal, e.g., a number of measurements, sampled at the sampling rate, over some amount of time. This will be a series or vector of numbers corresponding to values of some signal, measured over time. What does the Fourier transform yield? That's the term to the right of the first equal sign—it yields another function X, but this time, this is a function of frequencies f. In other words, this will be a series of power values (we already plotted some in the bottom panel of Fig. 5.2), one per frequency (not time). That is what we want—we want to transport the signal from the time domain into the frequency domain. So far, so good. How do we get here? The term to the right of the next equal sign shows how. This term represents the (input) signal—represented by the function $x(t)$ again, in the time domain. The Greek capital letter sigma stands for summation or, in this case, integration, integrating the signal that is a function of time over all time, from minus infinity to plus infinity (i.e., a long time indeed). The dt indicates that we integrate little pieces of time (infinitesimally small increments, actually). What is left to explain is the role of the exponential term, with which the signal $x(t)$ is multiplied.

First, let us elaborate on the fact that the signal *is* being multiplied with something. The transform takes the inner-product (also known as the dot-product) between the signal in the time domain and the exponential term, much akin to when the signal in the time domain was multiplied element-wise with the kernel, when we introduced convolution at the end of the last chapter. This kind of arrangement can be used to filter the signal for certain properties and to detect their presence. The color-detecting cones in the retina can be conceptualized in this way—the incoming visual stream is passed through separate color filters that measure the degree to which a certain wavelength (color) is present in the input signal (Gegenfurtner & Kiper, 2003).

But what is this filter? What is the Fourier transform looking for, when it analyzes and filters the signal? *Sine-waves.* The Fourier transform is looking for the presence of sine waves with certain frequencies and outputs the degree (this is conceptualized as "power") to which these sine waves are present in the input signal.

This is far from obvious. For starters, there is no "sine" term visible anywhere in the exponential term, and it is not clear why anyone would want to do that—filtering the signal for the presence of sine waves—in the first place.

Rest assured that the sine wave is right there, in the exponential, but is in deep disguise. There are hints though: upon closer inspection, the exponent contains the terms 2, π, as well as f (frequency) and t (time). If this sounds familiar, it should. We recognize this as the representation of a sine wave—we took the sine of 2*pi*frequency*time above, when plotting sine waves in Figs. 5.1 and 5.2. But where is the *sine* (that we used above) that takes these arguments? This mystery is resolved by clarifying another one (one of the rare cases where this actually works), namely why e and i are involved and what i is. Perhaps we should start with the i, to finish defining our terms. i stands for the *imaginary* part of a complex number and is defined as the square root of -1. The square root of -1 is not defined in the realm of real numbers, which is why the solution lies beyond the real numbers, in the *complex* plane. This is a good point to introduce the concept of complex numbers, we'll need it later. In signal processing, a signal is often represented as complex numbers, where the imaginary part can be used to represent phase. See Fig. 5.6 for an illustration of complex numbers.

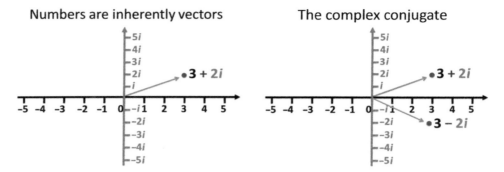

FIGURE 5.6   Complex numbers. Left panel: The complex plane. Real numbers are represented by the classical number line (here the x-axis, in black). All real numbers have no imaginary part. There is an additional axis (here, the y-axis, in red) that represents the imaginary part of complex numbers. Together, these axes span a complex plane. Whereas real numbers are scalars, complex numbers are inherently vectors with both real and imaginary parts. Right panel: The *complex conjugate*. It will be important later. It is simply the mirror image of the complex number across the number line (the real axis): **3 - 2i** is the "complex conjugate" of **3 + 2i**. A fancy term for a geometrically simple idea.

Taking a closer look, we recognize that all terms involved are also present in *Euler's formula*:

$$e^{ix} = \cos x + i \sin x \tag{5.2}$$

EQUATION 5.2   Euler's formula.

This is the key—Euler's formula established the relationship between *exponential* and *trigonometric* functions (such as sine and cosine), which allows sine waves to be represented in disguise here (Euler, 1748). Don't get discouraged if you don't get this the first time around. These things are complex, literally. We suggest sleep and rereading.

But say this is true—so what? Why filter the signal for the presence of sine waves?

Unless you are a mathematician, you are probably wondering what the big deal is about sine waves (likely since around middle school). Well, this isn't about sine waves per se. It is about describing the motion of going around a circle mathematically. There are several equivalent ways to describe circular motion, one that involves decomposing the

Cartesian coordinate on the circle into an *x*- (cosine) and a *y*-coordinate (sine) and one that involves moving on the radius itself, by the phase angle (angular distance), see Fig. 5.7.

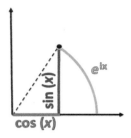

FIGURE 5.7 Moving around on a circle segment: The position on a circle described via Cartesian coordinates (red and blue) and via angular distance green end up at the same point, as we would expect from Euler's identity.

This also explains why sine waves can be described by three parameters—the amplitude (corresponding to the radius of the circle), the frequency (corresponding to how many times per second we go through the entire circle), and the phase (where on the circle we start to move). Again, sine waves simply describe circular motion. This will become more clear in Fig. 5.8.

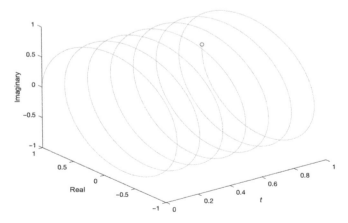

FIGURE 5.8 Tracing a complex sine wave.

By now, you have had to take a lot on faith, so this is perhaps a good point for demonstrations—and some coding. Importantly, let us convince ourselves that the exponential term in the Fourier transform really does represent complex sine waves, by plotting some.

If you are working through this chapter in one sitting and haven't cleared your workspace, this should work—we reuse some of the variables we defined above.

Pseudocode

---

1. Open a new figure
2. Making a trace, where exp stands for e
3. Show an animation of the trace, as it is being traced out
4. Add axis labels
5. Import new package
6. Specify 3d in Python

Python	MATLAB
| ```
from mpl_toolkits.mplot3d import Axes3D #5
fig = plt.figure() #1
ax = fig.gca(projection='3d') #6
trace = np.exp(1j*2*np.pi*freq2*t); #2
plt.xlabel('t'); plt.ylabel('real'); #4
ax.plot(t,np.real(trace),np.imag(trace))#3
``` | ```
figure %1
trace = exp(i*2*pi*freq2*t); %2
comet3(t, real(trace), imag(trace)); %3
xlabel('t'); ylabel('real'); zlabel('imaginary'); %4
``` |

This works out of the box. Good thing we didn't use "*i*" as an index or counter before. "*i*" and "*j*" are predefined in MATLAB. The plot of the trace should look something like Fig. 5.8. In Python, we express *i* by adding *j* to the value, as in: `np.exp(1j*2*np.pi*freq2*t)`.

Looks a lot like a spiral or slinky. If we put the plot on its ear by changing the perspective, we can get a deeper appreciation for sinusoidal as well as circular components of the trace: (Figs. 5.9 and 5.10)

Pseudocode

---

1. Find the handle of the trace in the plot
2. Make the trace black and thicker
3. Set azimuth and elevation to zero

Python	MATLAB
| ```
ax.plot(t,np.real(trace),np.imag(trace),c='k',lw=2) #2
ax.view_init(0, 90) #3, axes flipped in py
``` | ```
h = findobj(gcf,'type','animatedLine') %1
h(1).Color = 'k'; h(1).LineWidth = 2; %2
set(gca,'view',[0 0]) %3
``` |

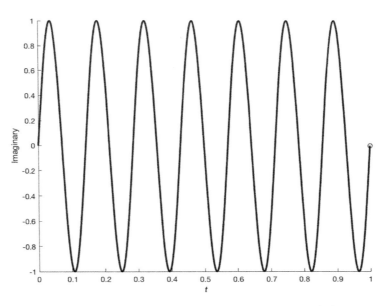

FIGURE 5.9 Looking at the trace in two dimensions—from the right vantage point—clearly reveals the sine wave. It was represented by the imaginary part of the exponential. We count seven cycles, which is what we put in. And all without ever calling the "sine" function.

Pseudocode
1. Change perspective to azimuth 90 and elevation zero
2. Make the axis square to avoid distortions from unequal aspect ratios

Python	MATLAB
ax.view_init(0, 0) #1	set(gca,'view',[90 0]) %1
plt.axis('square') #2	axis square %2

You can visualize this trace from any perspective—at will, by dragging and dropping—if you type `rotate3d` in MATLAB.

At this point, it would be hard to deny that this exponential term allows us to move on a circle or (if we include time as an axis) on a spiral.

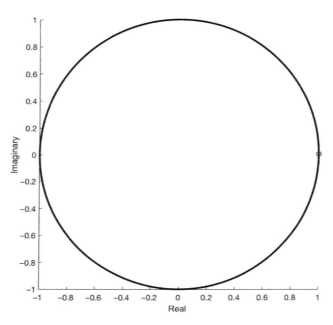

FIGURE 5.10   Collapsing over time by changing the view shows the overall circular structure of the trace, real vs. imaginary parts.

Why do circles matter? Because we can reconstruct any path or trajectory by adding enough circles—circles on top of circles or epicycles (Toomer, 1984).

This is usually more easily shown by using the *inverse Fourier transform*. We could, for instance, try to synthesize a square wave from adding adequately picked sine waves.

Properly aligned sine waves can, via *constructive* and *destructive* interference (alignment of peaks and troughs, see Glossary), recreate any signal, although to recreate a sharp edge without any ripples, we'll need to add an infinite number of them. That's impractical in practice, so there will be distortions or ripples, as you can see in Fig. 5.11—we just add three sine waves (in blue) and the resulting square wave (in black) is far from flat. More on this later.

This seems like a lot of work to devote to understand a single concept. We maintain that this is time well spent and predict that if you are seeing this for the first time and you continue in this field, you'll use Fourier transforms implicitly or explicitly for the rest of your career. It is a concept worth understanding thoroughly.

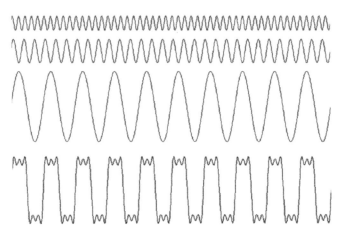

FIGURE 5.11 Synthesizing a square wave by adding suitable sine waves. Top blue wave: a. Middle blue wave: b. Bottom blue wave: c. Black trace: a+ b+ c, approximating a square wave.

The power of the Fourier transform lies in its generality. It can be (and is) applied to the analysis of a wide variety of signals, in neuroscience these include visual signals, sounds, EEG, LFP, and fMRI, to name just a few. However, it took well into the second half of the 20th century for the Fourier transform to be used so ubiquitously. The original Fourier transform is defined for analog signals, over all time. The signals we will want to analyze have been digitized and are typically varying over time—how the frequency content of a signal might change over time, e.g., in response to a stimulus, is exactly what will be of interest. The fast Fourier transform (FFT) algorithm introduced by Tukey allows us to do a *discrete Fourier transform* (i.e., on a digital, or nonanalog signal) on a short time window in which the frequency content is presumably stable and (as the name suggests) allows us to do this quickly and without needing an infinite number of sine waves.

Which brings us to the *spectrogram*, which is a visual representation of power at different frequencies over time. Conceptually, it corresponds to a stack of FFT outputs (like the ones we did to produce Fig. 5.3), stacked in time. But it isn't quite the output of the FFT itself that we are plotting. We are plotting power instead. Power is defined as the output of the FFT multiplied by its complex conjugate. An equivalent version of this is taking the absolute value of the squared FFT output, which also corresponds to power. Both operations get rid of phase information. This means that the spectrogram only uses the information that is contained in the amplitudes. A complementary plot would be a phase coherence plot (that uses only phase information), but that is "beyond 1" material—we have some references on that at the end of this chapter.

The spectrogram results from doing the FFT on the snippet of the signal that falls into a "window," plotting the frequency content in the window, then moving the window in time and plotting the frequency content again (and again) until the window has moved across the entire signal.

"The spectrogram" of a signal is somewhat of a misnomer. "A spectrogram" of a signal would be more apt. As you will see shortly, two spectrograms of the same signal can look dramatically different, depending on our choices of window and how far to move the window before doing the FFT again. Important window choices are *width* and *shape*.

But we are getting ahead of ourselves. Before we get lost in abstract discussions of signal processing concepts that mean nothing to you unless you have encountered them before, let us create a spectrogram of a signal we already have, then discuss these matters with specific code and figures (Fig. 5.12). The following code presumes that you have the signal processing toolbox installed in MATLAB. In Python, we will import the `scipy.signal` library.

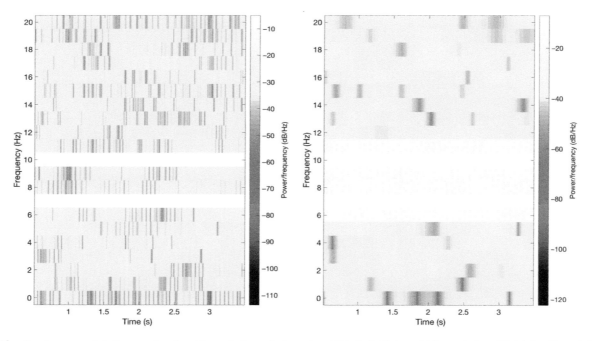

FIGURE 5.12 Spectrogram of stationary signals with a window of adequate width. Left: Using a Kaiser window. Right: Looking at the same data through a Hanning window.

Pseudocode	1. Define the number of elements in the window, here 1024
	2. Create a Kaiser window of that length
	3. Define that the overlap is maximal (1023 elements)
	4. Define the frequencies we want to assess and plot on the y-axis
	5. Open a new figure
	6. Open a new subplot
	7. Do the spectrogram with the right parameters. In addition to the ones we defined, we also supply the sampling rate and the parameter "yaxis" to put it into the format we want (frequencies on the y-axis, time on the a-axis) (default to –1 axis in Python)
	8. Define a new window of the same length, a "Hanning" window
	9. Open a new subplot
	10. Do the spectrogram on the same signal, with the same parameters, but a different window
	11. Change colormap to hot
	12. Show graph
	13. Import Python Package
	14. Plot colormesh
	15. Label axis
	16. Set y-limits

Python
```
import scipy.signal as sg #13
windLength = 1024; #1
wind = np.kaiser(windLength,0); #2
overl = len(wind)-1; #3
yFreqs = range(21); #4
fig = plt.figure() #5
plt.subplot(1,2,1) #6
f, tt, Sxx =sg.spectrogram(signalAndNoise,fs,wind,len(wind),overl) #7
plt.pcolormesh(tt,f,Sxx,cmap='hot') #!4
plt.ylabel('Frequency (Hz)');plt.xlabel('Time (sec)') #15 label axes
plt.ylim([0,20]) #16
wind = np.hanning(windLength);#8
plt.subplot(1,2,2) #9
f, tt, Sxx =sg.spectrogram(signalAndNoise,fs,wind,len(wind),overl) #7
plt.pcolormesh(tt,f,Sxx,cmap='hot') #14
plt.ylabel('Frequency (Hz)');plt.xlabel('Time (sec)')#15 label axes
plt.ylim([0,20]) #16
```

MATLAB
```
windLength = 1024; %1
wind = kaiser(windLength); %2
overl = length(wind)-1; %3
yFreqs = 0:20; %4
figure %5
subplot(1,2,1) %6
spectrogram(signalAndNoise,wind,overl,yFreqs,fs,'yaxis'); %7
wind = hanning(windLength Switch; and)%8
subplot(1,2,2)%9
spectrogram(signalAndNoise,wind,overl,yFreqs,fs,'yaxis'); %10
colormap(hot) %11
shg %12
```

This might be one of the few times where using a *Kaiser* window is actually superior. It allows us to clearly resolve the two frequencies we know to be present (we made the signal when creating Fig. 5.3), whereas the two frequencies are kind of blending together when looking at the same signal through the *Hanning* window.

Different windowing functions emphasize different things and differ in how much and how fast they attenuate *side-lobes* of power and how much they distort the signal. The need for using a proper window arises from the fact that the Fourier transform itself is defined for infinite time. If we split the signal into pieces and look at each piece separately, this will introduce *edge artifacts*, similar to the edge effects seen with convolutions in Chapter 4, Correlating Spike Trains. The name of the game in windowing is to pick the right window to avoid these edge artifacts as much as possible. Recall that it is hard to create sharp edges with a finite number of sine waves. In other words, a sharp window will cause rippling in frequency space (the Fourier transform of a square wave [or boxcar] function is a *sinc* function). So there are various windows that try to amplify the center of the window, but arrive at the edge in a smooth fashion. Again, we recognize that this is rather abstract. In MATLAB, you can simply look at the windows and their characteristics by typing `wvtool(wind)` (see Fig. 5.13).

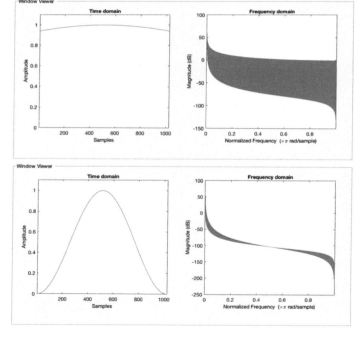

FIGURE 5.13 Top: The Kaiser window we used. Bottom: The Hanning window we used. Left panels: The window characteristics in the time domain. Note that the Hanning window goes to zero, but the Kaiser window does not. Right panels: The window characteristics in the frequency domain. The Kaiser window produces a lot of rippling in the side-lobes, which is why we don't usually want to use it. You can see this rippling in the noise of Fig. 5.12. The noise in the random part of the signal (outside of frequencies 7 and 10) appears much more "jumpy" when looking through a Kaiser window, whereas the Hanning window seems to smooth and calm things down quite a bit outside of the signal. The price to pay for this is that it blends the signal a bit more—7 and 10 Hz are harder to resolve.

Generally speaking, you'll want to use window lengths that are powers of 2 - (2, 4, 8, 16, 32, 64, etc.) because they are faster to compute, for algorithmic reasons in the FFT. Here, we use a rather wide window with 1024 elements. But we are looking at a signal that doesn't change over time. We know this because we created the signal ourselves to make Fig. 5.3. We also know from Fig. 5.3 (bottom panel) that if we use the entire signal of 4001 elements, we can resolve that there is a narrow signal at 7 Hz and a signal at 10 Hz without a problem, even in the presence of considerable noise. But note that this is asking a lot—the Nyquist frequency is 500 Hz, so we are interested in minute differences in frequency. If we use a window that is too narrow, the noise will prevent us from being able to resolve this difference. Even halving the window length will be too drastic to resolve these frequencies in the Hanning window, and even the Kaiser window is starting to have trouble. Run the code above again, but set windLength to 512, to yield Fig. 5.14.

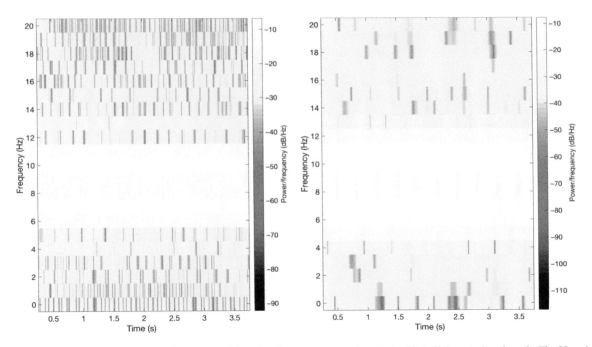

FIGURE 5.14 Spectrogram of the same signals as in Fig. 5.9, using the same parameters, but with half the window length. The Hanning window is starting to fail to resolve the difference between the two signal frequencies (7 and 10 Hz).

Generally speaking, there is a fundamental tradeoff (a kind of uncertainty principle) between determining the frequency content of a signal and determining when this frequency content occurred. The longer a window is, the better the estimate of the frequencies it contains. There are more data to go around, more noise to cancel out and so on. The price for this is a long window, so it is unclear when any of this happened (in the limit, we use a single window for the entire signal, as in the simple FFT, where we have no idea when any of this happened). In contrast, a very narrow window will be quite precise about when the frequencies happened, but not which ones. Any spectrogram will have to make a reasonable choice between these two extremes. Doing this in a principled fashion is a challenge and an active area of ongoing research (Casey, 2012). An alternative approach is not to commit to any particular window, but simply do multiple ones at the same time, then average them, usually using Slepian functions (Mitra, 2007). This "multitaper" approach is reasonable, but in practice may yield spectrograms that appear a bit "blobby."

So far, we have only considered spectrograms of *stationary* signals, in order to understand what spectrograms do, but the point of doing spectrograms is to look at signals that vary in frequency over time—if the signal doesn't change, you can just do a single Fourier transform over the entire signal. So let us move on to spectrograms of signals where the frequency content changes over time—if the signal doesn't change, you can just do a single Fourier transform over the entire signal. One of the simplest such signals is a "chirp," which has recently gained tremendous scientific publicity, as it seems to be the signal emitted from colliding black holes, relevant for the detection of gravity waves (Abbott et al., 2016).

Pseudocode	
	1. Create a new time base, a 2-second signal
	2. Create a chirp over that time, one that starts at 100 Hz, crosses 200 Hz at 1 second and is quadratic
	3. Listen to it
	4. Open a new figure
	5. Open a new subplot
	6. Plot amplitude of the signal over time
	7. Open a new subplot
	8. Set window length to 128
	9. Set overlap to 127
	10. Determine that we want to show 250 frequency bins on the y axis
	11. Do a spectrogram of the signal with a 128 element Kaiser window
	12. Open a new subplot
	13. Do a spectrogram of the signal with a 128 element Hanning window
	14. Open a new subplot
	15. Do a spectrogram of the signal with a 128 element Chebichev window

(*Continued*)

Python	
| Python | ```python
time = np.linspace(0,2,fs*2)#1
y=sg.chirp(time,100,1,200,'quadratic'); #2
#3 playing sounds is beyond "1" in Python
fig=plt.figure(figsize=(10,10))#4
ax = plt.subplot(4,1,1)#5
ax.plot(time,y) #6
ax = plt.subplot(4,1,2) #7
windLength = 128; #8
overl = windLength-1; #9
freqBins = 250; #10
wind=np.kaiser(windLength,0)
f, tt, Sxx =sg.spectrogram(y,fs,wind,len(wind),overl); #7
plt.pcolormesh(tt,f,Sxx);
ax = plt.subplot(4,1,3) #12
wind=np.hanning(windLength);
f, tt, Sxx =sg.spectrogram(y,fs,wind,len(wind),overl); #7
plt.pcolormesh(tt,f,Sxx)
ax = plt.subplot(4,1,4); #14
wind=sg.chebwin(windLength, at=100);
f, tt, Sxx =sg.spectrogram(y,fs,wind,len(wind)); #7
plt.pcolormesh(tt,f,Sxx);
``` |
| MATLAB | ```matlab
time = 0:1/fs:2; %1
y=chirp(time,100,1,200,'quadratic'); %2
sound(y,fs)%3
figure %4
subplot(4,1,1) %5
plot(time,y) %6
subplot(4,1,2) %7
windLength = 128; %8
overl = windLength-1; %9
freqBins = 250; %10
spectrogram(y,kaiser(windLength),overl,freqBins,fs,'yaxis'); %11
subplot(4,1,3) %12
spectrogram(y,hanning(windLength),overl,freqBins,fs,'yaxis'); %13
subplot(4,1,4) %14
spectrogram(y,chebwin(windLength),overl,freqBins,fs,'yaxis'); %15
``` |

This code will yield Fig. 5.15.

Some notes: Fig. 5.15 nicely illustrates the benefits of looking at a spectrogram. It is obvious that the frequency starts at 100, then is rising over time, in a quadratic fashion, which corresponds to what we hear when we listen to the signal.

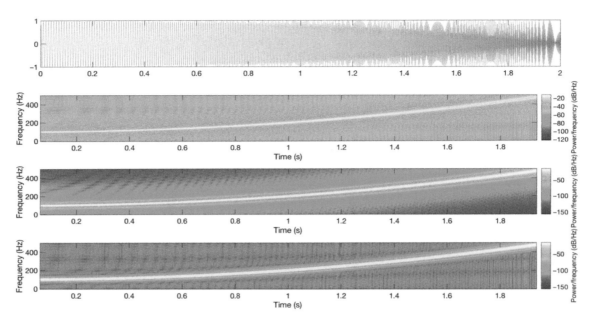

FIGURE 5.15 Looking at the same signal—a time varying quadratic chirp in the time domain (top panel), through a Kaiser window (second panel), a Hanning window (third panel) and a Chebichev window (bottom panel). Again, note the power of looking at power (over time). Powerful. Brave.

Looking at the same signal in the time domain (top panel) is hopeless. The Chebichev window looks a lot like a Gaussian (use the `wvtool` to confirm this) and seems to perform best here—with a strong suppression of the power in the noise and a clearly defined signal. The results of doing the signal processing with a Kaiser window are afflicted by brutal noise, which is characteristic of the Kaiser window—it usually produces severe rippling in the side lobes. The Hanning window performs somewhere in between. Instead of specifying which frequencies we want to plot on the y-axis, we specified that we want 250 divisions (which will correspond to 2 Hz bins, given the Nyquist frequency).

So far, we have always used almost complete window overlap, i.e., an overlap of 511 elements if the window length was 512 or an overlap of 127 if the window length was 128. This leads to plots with the most resolution, but is computationally the most expensive. If you are in a rush, use less overlap. This will lead to blocky plots, but they are done much faster. For an extreme case of this, see Fig. 5.16—we use the same signals and do spectrograms with the same windows, but no overlap at all. This figure is produced by setting `overl` to 0, but otherwise running the same code that produced Fig. 5.15:

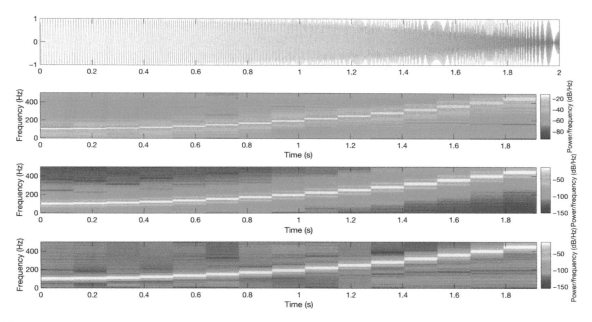

FIGURE 5.16 Same signal and windows, but no overlap. As you can see, the spectrogram is now very "blocky." The signal is simply divided into pieces of 128 elements and the fft is computed on those elements, then the window is advanced by an entire 128 elements and the fft is done again. That's what yields the discontinuities between windows. Generally speaking, spectrograms with more overlap will appear much smoother, but take a lot longer to compute.

This is a good time to introduce the notion of *filtering*. Understand that we can only scratch the very surface of this here—explaining why it is done, what it is, and doing a simple example of it. For a more comprehensive treatment of filtering, see Bibliography. We want to emphasize this because there is no way we can do the topic justice on a few pages (which is all we have here), people get entire degrees in this kind of thing. None of it is obvious or intuitive, all of it is complicated.

Filtering of signals also occurs in nature, more or less explicitly. When lightning strikes far away and you hear the rumbling thunder some time later, the atmosphere has effectively performed a low-pass filtering of the signal, emphasizing lower frequencies as they carry farther in air—higher frequencies lose power much faster. The uterus similarly acts as a low-pass filter—strongly attenuating frequencies above 300 Hz—keep that in mind next time you sing or talk to your fetus (Querleu, Renard, Versyp, Paris-Delrue, & Crèpin, 1988; Abrams et al., 1998; Gómez et al., 2014). When you use an edge detection algorithm in Photoshop, you are running a high-pass filter on the image, under the hood. When you record data with your electrophysiology rig and don't want to be swamped by line noise, you'll need to use a band-reject filter centered on 60 Hz (in the United States; 50 Hz in Europe).

So it is worth understanding what is going on. We'll re-use the signals we already have, but now create versions that have been filtered in different ways.

There are many filter types in use in electrical engineering and signal processing. The one we will cover here is the one we think you are most likely to use in practice, namely the *Butterworth* filter. It has several nice characteristics, including a flat frequency response in the *passband* (the region of frequencies you wish to not get filtered out), with power gently falling off outside of it—as smooth as butter, without introducing distortions or ripples (Butterworth, 1930). The ideal filter would completely pass everything in the passband (i.e., allow through the parts of the signal we care about) and completely reject everything outside of it, but this cannot be achieved in reality—the Butterworth filter is a close approximation. Other filters, such as the Chebishev filter, reject frequencies outside of the passband much more sharply, but this comes at the price of ripples in the passband itself, which is often unacceptable.

Let's create four filters, two Butterworth and two Chebishev, to illustrate this. Filters are defined by their filter coefficients (usually called "A" and "B"). In the code below, we first make a fifth order, then 10th order filters. In software, the order of a filter pertains to the number of polynomials that are involved, in hardware the number of resistors and capacitors.

Pseudocode	1. Create a fifth order Butterworth low-pass filter that passes the lowest ⅓ of the frequency range below Nyquist (1 would be the Nyquist frequency)
	2. Create a 10th order Butterworth low-pass filter that passes the lowest ⅓ of the frequency range below Nyquist
	3. Create a fifth order (type I) Chebichev low-pass filter that passes the lowest ⅓ of the frequency range below Nyquist and accepts 7 dB of ripple
	4. Create a 10th order (type I) Chebichev low-pass filter that passes the lowest ⅓ of the frequency range below Nyquist and accepts 7 dB ripple
	5. Python code to approximate MATLAB's `fvtool` function
Python	``` b1_low, a1_low = sg.butter(5, .2, 'low', analog=True)#1 b2_low, a2_low = sg.butter(10, .2, 'low', analog=True)#2 b3_low, a3_low = sg.cheby1(5, .2, 100, 'low', analog=True)#3 b4_low, a4_low = sg.cheby1(5, .2, 100, 'low', analog=True)#4 w, k = sg.freqs(b3_low, a3_low) #5 plt.semilogx(w, 20 * np.log(abs(k))) #5 ```
MATLAB	``` [B1_low,A1_low] = butter(5,0.2,'low'); %1 [B2_low,A2_low] = butter(10,0.2,'low'); %2 [B3_low,A3_low] = cheby1(5,7,0.2,'low'); %3 [B4_low,A4_low] = cheby1(10,7,0.2,'low'); %4 ```

Note that the Chebichev filter requires four parameters—that's because we have to specify what degree of ripple (in dB) in the passband is acceptable. The more ripple we deem acceptable, the more sharply the Chebichev filter is able to suppress frequencies outside of the passband. As the Butterworth filter is maximally flat in the passband, we don't need to specify that parameter for Butterworth. Fig. 5.17 was created by looking at the filter characteristics with the filter visualization tool, typing `fvtool(B,A)`, where B and A stand for the suitable coefficient from the 8 above. Note that this tool plots the y-axis in log scale, so the falloff of the Chebichev filter is more dramatic than it looks.

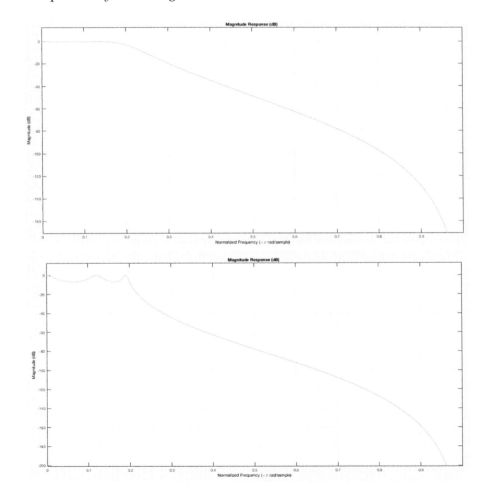

FIGURE 5.17 Filter characteristics of fifth order (top panel) and 10th order (bottom panel). Butterworth (left) and Chebichev I (right) filters.

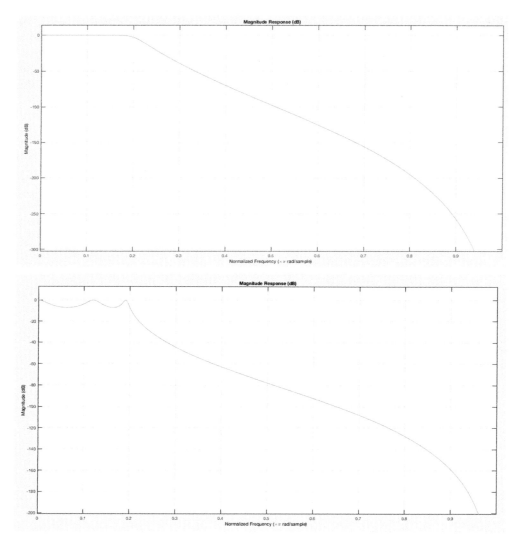

FIGURE 5.17   *(Continued)*

As you can see, the lower-order Butterworth filter starts attenuating even before the end of the passband and falls off rather gradually. The higher-order Butterworth filter starts attenuating much closer to the passband and drops off steeper. The Chebichev filters ripple along but achieve more precipitous rejection outside of the passband than the Butterworths. Which one you want to use in practice depends on your use case.

For now, let's adopt the eighth order Butterworth filter in order to do some filtering of the signal.

If you listen to this and compare it to the original chirp, filtering with Butterworth simply corresponds to an attenuation of amplitude, the frequency content of the sound is not distorted.

Pseudocode	1. Create an eighth order Butterworth low-pass filter with a cutoff at 0.6 Nyquist 2. Create an eighth order Butterworth high-pass filter with a cutoff at 0.4 Nyquist 3. Filter the chirp signal with the low-pass filter 4. Filter the chirp signal with the high-pass filter 5. Filter the filtered signal we created in (4) again with a low-pass filter to create a band-pass filtered version of the signal
Python	``` B_low,A_low = sg.butter(8,0.6,btype='low',analog=True) #1 B_high,A_high = sg.butter(8,0.4,btype='high',analog=True) #2 winds = {} winds['y'] = y winds['yLow'] = sg.filtfilt(B_low,A_low,y); #3 winds['yHigh'] = sg.filtfilt(B_high,A_high,y);# %4 winds['yBand'] = sg.filtfilt(B_low,A_low,winds['yHigh']);#5 ```
MATLAB	``` [B_low,A_low] = butter(8,0.6,'low') %1 [B_high,A_high] = butter(8,0.4,'high') %2 yLow = filtfilt(B_low,A_low,y); %3 yHigh = filtfilt(B_high,A_high,y); %4 yBand = filtfilt(B_low,A_low,yHigh); %5 ```

Note that simple filtering would cause a phase-shift of the filtered signal. We don't want that, so we filter twice, once forward and once backward, in order to yield a filtered, but not phase-shifted version of the signal. In MATLAB and scipy.signal this is achieved by the function filtfilt.

Now, let's look at and listen to these filtered signals and compare to the original (Fig. 5.18).

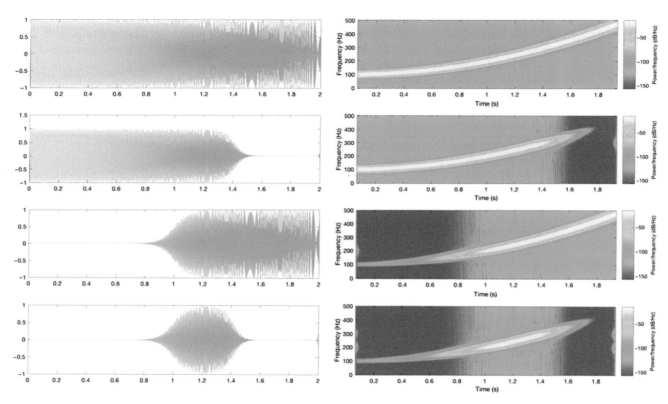

FIGURE 5.18 The effects of filtering. Top panel: original chirp. Second panel: low pass. Third panel: high pass. Bottom panel: band-pass. Left-panels: Signal as amplitude over time. Right panels: Spectrograms.

Pseudocode	
	1. Open figure
	2. Higher overlap, so that spectrograms are smooth again
	3. Define a cell of four signals: The chirp and filtered versions thereof
	4. Initialize a counter
	5. Open the subplot that corresponds to the counter number
	6. Check if the counter is odd or even with the mod function
	7. If it is odd, plot the amplitude of the suitable signal over time and increment the counter
	8. If it is even, plot the spectrogram of the suitable signal, play the sound that correspond to it and wait for an appropriate length of time, then increment the counter (code not included in Python)
	9. Loop over column/types of plot
	10. Loop over rows/signals

Python

```python
fig=plt.figure(figsize=(10,10))#1
overl = windLength-1; #2
signals = ['y','yLow','yHigh','yBand']; #3
counter = 1; #4
for ii in range(4): # 10
    for jj in range(2):# 9
        ax=plt.subplot(4,2,counter) # 5
        if counter%2 == 1: # 6
            ax.plot(time,winds[signals[ii]]) # 7
            counter = counter + 1; # 7
        else : #6
            f,tt,Sxx=sg.spectrogram(winds[signals[ii]],fs)
            plt.pcolormesh(tt,f,Sxx);
            counter = counter + 1; # 8
```

MATLAB

```matlab
figure %1
overl = windLength-1; %2
signals = {'y','yLow','yHigh','yBand'}; %3
counter = 1; %4
for ii = 1:4 %10
    for jj = 1:2 %9
        subplot(4,2,counter) %5
        if mod(counter,2) == 1 %6
            plot(time,eval(signals{ii})) %7
            counter = counter + 1; %7
        else    %6
spectrogram(eval(signals{ii}),chebwin(windLength),overl,freqBins,fs,'yaxis'); %8
            sound(eval(signals{ii}),fs) %8
            pause(length(eval(signals{ii}))/fs) %8
            counter = counter + 1; %8

        end %6
    end %9
end %10
```

Note that the Butterworth filter is so smooth that we never quite cut the power to zero on the low end. If you feel adventurous, try a higher-order Butterworth filter or a Chebichev filter and see if you can get it to zero. Also note that we implemented the MATLAB code in a nested loop here. Per "eval," we modified what is plotted per iteration, instead of writing things out four times.

That's it for signal processing concepts. Let us apply what we learned to real data. In this case, local field potentials (LFPs) that were recorded in the auditory cortex in response to sound signals. LFPs are the lower-frequency components in the voltage signal. If spikes have most of the power above 1000 Hz, LFPs are defined as the spectral component of the voltage signal below 1000 Hz, usually well below that—with most of the action below 100 Hz (anything above that is considered to be *high gamma*). Whereas voltage spikes correspond to the outputs of single neurons, LFPs are thought to reflect the synchronized inputs to large populations of neurons, as they reflect *EPSPs* and *IPSPs* in their apical dendrites (Logothetis, Pauls, Augath, Trinath, & Oeltermann, 2001). How "local" the local field potential is, is subject to active debate (Kajikawa & Schroeder, 2011).

Thus far, we have built all the programs from the bottom up, step by step. From now on, some of the more advanced programs in each chapter will be too complex to do so effectively, as we are drawing from many concepts and building blocks. Instead, we will provide you with programs we wrote, programs that are heavily commented. It is useful to try to reverse-engineer them.

The program to do the LFP analysis can be downloaded from our companion website in POM. Here is what the program does, in brief.

First it loads the workspace `"mouseLFP.mat."` The data in the file were recorded in four electrophysiological recording sessions from mouse auditory cortex, one session for each row in the DATA cell. The experiment itself consisted of presenting 200 auditory stimuli (each presented for 50 ms, with a 500 ms interstimulus interval, sampled from two tones, in random order). Data were sampled at 10 kHz. The seven columns represent (in order): (1) Voltage snippets per trial (tone-locked). (2) Background noise per trial (outside of stimulation). (3) Raw voltage trace over the entire session (not cut up). (4) Trace of triggers over the entire session (when tone was on, 1 if on, 0 if not). (5) List of tones, in order presented. (6) Tone on- and offset-time over the entire session. (7) Recording site, date, etc.

We then filter the signal with a low-pass Butterworth filter with a cutoff frequency at 1000 Hz. This is necessary because the DATA file contains the all-pass, broadband raw voltages. This includes spike information, but we want to look at the LFP alone, which is classically defined as being below 1000 Hz. We will use this filtered signal for analysis and plotting. This is a classical instance of the "filtering/cleaning" stage of the canonical processing cascade. The program then makes one figure per recording session, for a total of four figures. Each figure contains four subplots: two subplots for the ERP for each tone (including an error band), and two subplots for the corresponding spectrograms (one for each tone). We plot frequencies from 0 to 200 with a 5 Hz bin-width.

See Fig. 5.19 for one of the four figures that this program produces.

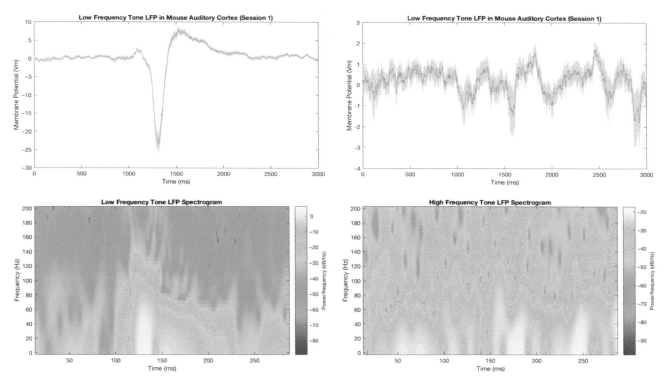

FIGURE 5.19 Plots from recording session 1. Top row: Event-related potentials (ERPs). Gray error bands represent the standard of the mean (SEM). Bottom row: Spectrograms. One way to conceptualize this is to think of them as the "singing of the heavenly choirs." ⌘Each chorister sings at a particular frequency, e.g., 100 Hz. The more yellow the trace is at a given time, the louder that frequency is sung at that time. The linear combination of all these tracks produces the signal that you can see (or hear, if you listen to the data at the sampling rate, which we recommend). Left column: Response to the low tone. Right column: Response to the high tone. As you can see, the neuron is tuned—there is a brisk response to the low, but not the high, tone in the LFP.

That's it. We believe we got you from 0 to 1 in signal processing of analog signals. To go beyond that and learn about such exciting topics as the Morlet wavelet, the filter Hilbert method, phase coherence, and multitaper methods, we warmly recommend a close read of Mike X Cohen's "Analyzing neural time series data" (2014).

⌘. *Pensee on harmonic oscillators in the brain:* We have noted in a previous chapter that it is fundamentally inappropriate to express spike rates in units of Hz. A power analysis of EEG, MEG, and LFP data is also somewhat puzzling. For instance, we determine which harmonic oscillators are most active within an EEG trace, e.g., those oscillating around 10 Hz if most of the power was in the alpha band or around 6 Hz if most of the power was in the theta band, *if there were harmonic oscillators.* To our knowledge, there are no neurons in the cortex that spike in any way that could meaningfully be described as purely sinusoidal. If there are any in the brain, they might be situated in subcortical structures like the suprachiasmatic nucleus—a small population of neurons that control circadian rhythms that discharge at a very low frequency—about 1 cycle a day (Buhr et al., 2010) or the olfactory bulb, a key structure in odor perception (Rojas-Libano & Kay, 2008). There is no question that neurons exchange information in the form of spikes, but the theoretical significance of these oscillations is extremely controversial, with positions ranging from high-powered signals in EEG and LFP dominating what happens in the brain, as they reflect synchronized activity that readily propagates through cortex (Fries, 2005, 2009) to the position that these are simply signals we happen to be able to measure conveniently and noninvasively, but that are fundamentally epiphenomenal in nature and have no role in the way neural populations communicate (Shadlen & Movshon, 1999). It is fair to say that the functional role of these oscillations (and their neural correlate—networks?) is still very much in question and subject to active research debate (Salinas & Sejnowski, 2001; Buszaki and Draguhn, 2004; Uhlhaas et al., 2009), far beyond this brief note and indeed the entire book.

CHAPTER

6

Biophysical Modeling

This is a pivotal chapter. So far, we have taken the data at face value and focused on their analysis. Starting in this chapter, we will begin to transcend the data—first by curve fitting, then (in subsequent chapters) becoming increasingly more abstract, using the data to predict, changing the dimensionality of the data, and looking for clusters in highly abstract conceptual spaces.

Moreover, in the past three chapters, we analyzed data from progressively larger ensembles of neurons. We first looked at one neuron in Chapter 3, Wrangling Spikes Trains, then pairs of neurons in Chapter 4, Correlating Spike Trains, and then aggregated signals from small populations of neurons in Chapter 5, Analog Signals. Here, we circle back to take a closer look at a single neuron. Even a single neuron is a world in itself, quite capable of performing some rather sophisticated computations to do information processing (Koch, 2004). How it is able to do so will become clearer when we look at the *biophysics* of the neuron. We will use this as an opportunity to *model* some of the relevant computational machinery. Like other tissue of the body, neurons are mostly made up of proteins and fat (much like a steak) although the content of omega-3 fatty acids, docosahexaenoic acid in particular, is far higher in the brain (Green & Yavin, 1998). Neurons are similar in a lot of ways to other cells in the body—they have DNA in their nuclei, are filled with cytoplasm, and contain the same organelles such as ribosomes, mitochondria, vesicles, and other organelles, just like muscle, skin, and bone cells.

Many cells communicate with cells in their neighborhood through means of electrical or chemical exchange. What sets neurons apart is the *channels* in their *cell membrane* which facilitate electrical communication with fellow neurons. Other parts of the neuron contribute to this as well, such as the axon hillock to generate spikes, a dendritic arbor to receive electrical information, and synapses to bridge the gap between neurons chemically. In this chapter, we will focus on the biophysical properties that underlie the *electrical* behavior of neurons. This will allow us to more deeply understand what spikes are and how they are generated. In previous chapters, we have simply accepted that neurons produce them and that we can measure the rate of their occurrence. Here, we wonder how the neuron does that: how does the neuron generate spikes?

Neural Data Science.
DOI: http://dx.doi.org/10.1016/B978-0-12-804043-0.00006-4

BIOPHYSICAL PROPERTIES OF NEURONS

A neuron in the *resting state* is a neuron that is not generating action potentials at that time. In this resting state, the voltage across the membrane of the neuron hovers around some characteristic value germane to the neuron. A typical *resting potential* (represented as vR in the code, below) for neurons in vivo is –60 millivolts (mV). Importantly, it is negative—indicating that there are more negatively charged ions inside the neuron than outside of it (Fig. 6.1). The existence of this voltage differential is owed to the fact that the cell membrane acts as a *capacitor*, separating (net!) positively charged ions on the outside from (net!) negatively charged ions on the inside. The amount of charge that a membrane can hold is its *capacitance* (represented as C in the code, below). As neurons have the ability to release this charge, they are similar to a battery in this regard.[96]

The ability for a neuron to change its voltage is due to many features of the cell: the *leak conductance* (represented as gL in the code) represents how much the negative charge happens to spill out of the cell if not at its resting voltage. This is a *passive* feature of the neuron, as no part of the neuron's machinery is *actively* leaking ions. We can think of the leak conductance as small holes in the membrane that happen to let some ions through (Fig. 6.2).

The *spike threshold* (represented as vT in the code) of a neuron is represented as a single value, usually something like –50 mV. As the *resting voltage* of the neuron is below the spike threshold, then, when at rest, by definition the neuron is

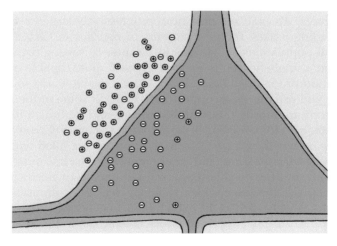

FIGURE 6.1 Ions of positive and negative charge sit around the outer member of the cell. The membrane of the neuron here acts as a capacitor to maintain this charge. This is an exemplary resting neuron: no ions moving, everything resting on either side of the membrane. You will be hard pressed to find such a peaceful cell in the wild.

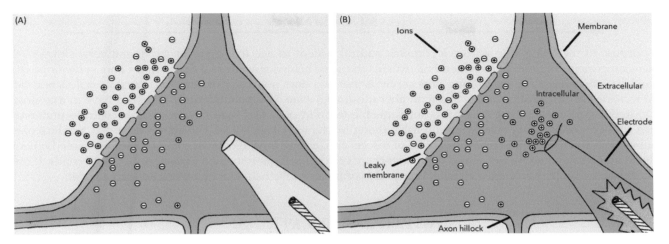

FIGURE 6.2 Left, electrode placed onto membrane, not broken into cell yet. Right, electrode breaks into membrane so that solution inside electrode is continuous with intracellular media. This neuron now has a membrane with holes in it, where ions can pass through. This embodies the notion of the leak conductance. Also, we've drawn in a glass electrode whose open tip is shown to have broken through the membrane so the internal solution in the electrode is continuous with the internal solution of the neuron. This apparatus permits us, the experimenters, to *inject* a current. Injection of a *positive* current will cause the neuron's voltage to become more positive, and move toward the spike threshold. If the spike threshold voltage is reached, the *axon hillock* will trigger an action potential, which will travel down the axon.

not spiking. A resting neuron can be considered to be silent. However if the voltage is hovering around −60 mV, and an external force "pushes" the voltage to the spike threshold, then, the neuron's voltage will jump rapidly to a high value, up to near +20 mV, and it will rapidly come back down to its resting voltage. This *spike* is a *high-frequency* event, as it happens very quickly: from start to finish a spike takes only about a millisecond.

How do we determine how much the membrane voltage changes when we inject current? For a given point in time, the instantaneous change in voltage (dV) follows Eq. (6.1):

$$dV = \left(gL * (V - vR) + I\right)/C \qquad (6.1)$$

where dV is the change in voltage, gL is the leak conductance, V is the membrane potential, vR is the resting membrane potential, I is the injected current, and C is the membrane capacitance.

MODELING

The purpose of modeling is to reduce a complex natural system to one that involves (comparatively) simpler—usually artificial—components in order to see whether it exhibits the same behavior as the real-life complex one, in the hope of uncovering causal relationships. The underlying philosophy is that one does not really understand a system unless one can build one that recreates its behavior or brings about another (desirable) state. What confidence would you have in a company that claims to understand how an engine works, but is unable to build one? Did man have a credible claim to understand the principles of flight before being able to build reliable flying machines (Jakab, 2014)? The latter example is a critical one—a lot of organisms can fly: birds, some mammals, a lot of insects and, a long time ago, even some reptiles. It seemed obvious that wings were involved somehow, but which components are critical for flight remained unclear until very recently. If one has a model system, one can play around with what is critical to bring about the behavior of interest and what is not.

WHY USE SIMULATIONS?

Neurons are cells that communicate with other neurons through electrochemical means. What triggers the chemical signal exchange at the synapse is the arrival of an *action potential*, which is an all-or-none or binary event. Thus, it is tempting to represent the presence of such an event with a 1, and the absence of the event by 0's. Whereas it might be permissible to reduce the behavior of a neuron to outputting 1's and 0's, it is useful to simulate some of the biophysical components that lead to this spiking activity. Other simulation endeavors are more ambitious. The European "Blue Brain Project" strives to simulate the activity of millions or more neurons using complex models in an attempt to recreate the behavior of a cortical column and ultimately the entire brain (Markram, 2006, 2012). The issue with attempting simulations of such complexity is that there are tremendous gaps in our understanding of the system, both of single neurons as well as networks of neurons. These gaps are filled with assumptions, stacked upon assumptions—i.e., assumptions about single neurons as well as assumptions on how neurons are connected locally as well as the connections between larger ensembles of neurons. The problem with a simulation that includes assumptions all the way down is that even if the probability that each assumption is almost one, the joint probability that they are all true at the same time is almost zero. If the simulation then fails to produce realistic behavior, involving so many nested assumptions makes it nearly impossible to tell which one is true and which one is false. In other words, the prospects of learning something from this are perhaps somewhat limited (Fregnac & Laurent, 2014). Models and simulations are most powerful when they are simple, not complex. This leads to challenges when attempting to simulate something as inherently complex as the brain. The temptation to do so is understandable, given how interesting such a complex system is. The point of simulations is to further our understanding. At the same time, one cannot meaningfully simulate what one does not understand sufficiently.

In contrast, the simulation of single neuron behavior is grounded in a rich biophysical literature. Here, it is useful for demonstrating how simple biophysical properties of neurons (in particular, resting and threshold voltages, capacitance, and membrane leakage) can account for and bring about a few basic behaviors of neurons, such as spiking. Specifically, we will simulate spike patterns of neurons for currents that have been "injected" into these neurons, and visualize how the amount of current injected over time can induce different spike patterns (Brette & Gerstner, 2005; Levy & Reyes, 2011). At the end, we will do some simple modeling of the simulated responses of neurons to injected currents. Along the way, we'll be introducing the concept of *objects* and *classes* in Python and MATLAB.

WHY OBJECT-ORIENTED PROGRAMMING?

In this chapter, we will scratch just the surface of *objects* in programming (0–1, like with everything else). If you dive deeper into Python, you'll eventually learn that *everything* is an object (!), but that is far beyond the scope of this book. MATLAB is also becoming increasingly more object-oriented. Our modest goal here is to simulate how a neuron responds to different current injections using some explicit *object-oriented programming* (OOP) methods. We could have just as easily developed these examples without taking this approach at all, but as a primary goal of this book is to get you, the reader, to become *code-safe*, we thought it necessary to provide some basic examples of what to expect when you encounter OOP-based approaches (of which you will find many in industrial software tools and increasingly also in academic software).

So, what is an *object*? An object is a one-stop shop of data and tools to do stuff with the data. Unlike a dictionary or list (or struct or cell), which can hold all kinds of data of different types, an object can also have its very own functions that come with it. An object can contain an array, functions that do calculations on the array, and plotting functions to visualize the results of these computations. An object can contain a bunch of *scalars* while also containing functions that act as equations for these variables. In this way, objects provide an easy way to *pass* around a bundle of data and tools all in one. The name of the game is *encapsulation*—the idea being that an object is a self-contained modular unit onto itself (Leibniz, 1714), that might or might not interact with its environment—if it does, it does so through clearly defined interfaces.

We define objects in POM by declaring *classes*. A *class* is the code that specifies variables it contains ("*properties*") as well as their default values, and has associated functions ("methods"). To create an object, we have to call the *class*, and assign the output of the class to a variable. This is called *instantiating* a class, and gives us an object as an *instance* of the class. Just like you are an instance of the category human (presumably), with certain attributes and capabilities, an object is a specific instantiation of an abstract category that we call a class. The relationship between object and class is the same as between exemplar and category, e.g., your dog is a particular instantiation of the category "dog." Functions, much like the Python packages we've been importing, or MATLAB's built-in functions, can be written so that they work specifically within the scope of objects. Functions within objects in POM are known as *methods*.

PYTHON IS INHERENTLY OBJECT-ORIENTED: HOW DOES MATLAB IMPLEMENT THESE THINGS?

Increasingly well. MATLAB is introducing more and more object-oriented concepts—completely revamping their object-oriented framework in the 2008 and 2009 releases. By now, pretty much every OOP concept in Python has an analog in MATLAB. Technically, most of these things are implemented by a mix of structs (that represent objects) and functions (that represent methods). Since release 2014b, most figure elements are also objects. For instance, when plotting a line as `h=plot(x,y)`, one could set the properties of this line with, for instance, `set(h,'color','k')`, to make it black. Now, one can simply type `h.` and tab—complete will bring up a list of available methods of the line object. Generally speaking, MATLAB makes minor and incremental changes to the code base—how functions work, what they are called and so on in each release. However, over a decade or so, these cumulative changes add up to transformational change in the way MATLAB works—there are 20 individual releases per decade.

CREATING THE CLASS NEURON

To create a class in Python, we start by defining it with `class Neuron`. In Python, it is common convention to capitalize the first letter of a class. Inside (indented in) the class, we first declare the method `__init__()`, "passing" the variable `self`. There are a couple of important concepts to note here: the method `__init__` is automatically called whenever the class `Neuron` is instantiated. That is, later, when we create an object that is an instance of the class `Neuron`, rest assured that the first thing the class Neuron does is run the method `__init__`. The variable `self` is a reference to the class itself. Inside the method `__init__`, when we assign the attribute `self.C` to the value 0.281, an object that was instantiated as class Neuron (e.g., by typing `myFirstNeuron = Neuron()`) now has the attribute C associated with it. We could type `myFirstNeuron.C` and it would print out 0.281. Inside of the `__init__` method, we specify the attributes C (capacitance), `gL` (leak conductance), `vT` (spike threshold), and `vR` (resting potential). Each of the attributes becomes associated with any instance of the class Neuron, and are available to the other methods in Neuron. This is one example of how objects are an "everything but the kitchen sink" kind of approach.

We create the method create_injection_current, using as input the object self, which, just like with the __init__ method, is a way of passing the object to the method, so that we can make new attributes associated with the object. It's worth noting that we could have named self anything—we could have very well been passing around something named taco but it is common convention in Python for the self-referencing object to be named self. Inside this method create_injection_current, we create a simulated current injection named currentInj, by appending together 10 zeros with an array of values increasing from 0 to 0.99 in steps of 0.01. This creates a ramping effect, so that the current injected will gradually increase. The variable currentInj is set as an attribute of self so that we'll have access to it in other methods. We calculate the length of currentInj and hold on to this as the attribute self.T.

Next we create the method leaky_integrate_and_fire, again calling as input the object self. We create an array called timeseries, which will be used for plotting. We then create an array for the membrane potential of the simulated neuron that is length self.T, and equal to the resting membrane potential, self.vR. This is our way of initializing the voltage trace. Note that it is this voltage trace that we'll be plotting. We next enter a while loop. A while loop is similar to a for loop, in that it will continue to iterate through whatever is inside the loop so long as a condition is met (in this case, as long as ii is less than self.T-2). The key to an effective while loop is initializing a variable (ii) before the while statement, and increasing the variable while still inside the loop (ii+=1). If you do not add one to the variable inside the loop, the while statement will keep evaluating the statement over and over again, because ii will never *not* be less than self.T-2. This is known as an infinite loop—avoid making one. You'll notice later that we use values up to less than self.T-2, as our spikes are 2ms in duration: we won't want to try to access a part of the array with an index larger than the length of the array self.V (if this doesn't make sense, you can set the statement to while ii < self.T-1 to see the error). Once inside the loop, we calculate how the voltage changes as a function of both the injected current and of the intrinsic properties of the cell (i.e., the attributes assigned in __init__). We effectively implement Eq. (6.1) here, to determine the change in voltage over 1ms, dV. Once this change in voltage is determined, we set the voltage at the next point in time (V at time ii+1) equal to V at the present time (ii) plus dV. We check if the voltage has reached the threshold self.vT, and, if so, set the voltage equal to a whopping +20mV, the voltage that a spike reaches, and then set the voltage at the time point after that (ii+2) back to resting potential, vR. This constructs a spike, and we increase our time counter (ii+=1). If a spike is reached (or even if not) we increment the time counter (also using ii+=1), so that we keep iterating through everything in time. The output of this function is a new self.V with changes in voltage and spikes whenever the threshold is reached.

We create the method plot_neuron (again using as input self), which uses functions similar to what we have seen in previous chapters, to visualize (1) the injected current over time, and (2) the voltage trace over time.

Pseudocode

1. Import relevant packages
2. Declare the class called *Neuron*
3. Define method __init__, which is automatically instantiated when Neuron is created
4. Declare the variable for capacitance (an object attribute)
5. Declare the variable for leak conductance
6. Declare the variable for spike threshold
7. Declare the variable for resting voltage
8. Create method create_injection_current
9. Create an array to simulate a current to inject
10. Calculate the length of the injected current
11. Create method leaky_integrate_and_fire
12. Create the values for the timeseries
13. Set the voltage V equal to the resting voltage, vR
14. Initialize the index
15. For values less than the length of time minus 2.
16. Calculate the change in voltage, implementing Equation X,
17. Set the following index value to the present voltage plus dV
18. Check if spike threshold has been crossed
19. If so, set the voltage to positive 20 mV
20. Reset the voltage to the resting potential
21. Increment the index in time if threshold was triggered (since the spike is 2 time units in length)
22. Increment the index in time if threshold was not triggered
23. Create method plot_neuron
24. Create the figure
25. Add the first row of two subplots
26. Plot the currentInj variable vs the timeseries variable
27. Add the title and make italic
28. Add the y-label and make italic
29. Add the second of two subplots
30. Plot the voltage trace V vs the timeseries
31. Set the title for the second subplot and make italic
32. Set the xlabel for the second subplot
33. Set the ylabel for the second subplot
34. Adjust the subplots and labels so nothing overlaps
35. Display the figure to the screen (uncomment to perform this)
36. Save figure
97. In MATLAB, we need to specify and initialize properties (variables of the object class) as well as methods explicitly, in the properties and methods block
98. In MATLAB, we need an explicit class constructor
99. In MATLAB, all properties used by the object have to be declared up front, explicitly

Python	MATLAB

```python
import numpy as np # 1
import matplotlib.pyplot as plt

class Neuron(): # 2
    def __init__(self): #3
        self.C= .281 # 4
        self.gL = .030 # 5
        self.vR = -60.6 # 6
        self.vT = -50.4 # 7

    def create_injection_current(self): # 8
        self.currentInj =
np.append(np.zeros(10),np.arange(100)/100.) # 9
        self.T = len(self.currentInj) # 10

    def leaky_integrate_and_fire(self): # 11
        self.timeseries = np.linspace(0,self.T-1,self.T) # 12
        self.V = np.ones(self.T)*self.vR # 13
        ii=0 # initial the index counter # 14
        while ii < self.T-2: # 15
            dV = (-self.gL*(self.V[ii] - self.vR)+
self.currentInj[ii])/self.C # 16
            self.V[ii+1]=self.V[ii]+dV # 17
            if self.V[ii+1]>=self.vT: # 18
                self.V[ii+1]=20 # 19
                self.V[ii+2]=self.vR # 20
                ii+=1 # increment #21
            ii+=1 # Increment outside of the if statement # 22

    def plot_neuron(self): # 23
        fig = plt.figure() # 24
        ax = fig.add_subplot(211) # 25
        ax.plot(self.timeseries,self.currentInj,c='k') # 26
        ax.set_title('current injection',style='italic') # 27
        ax.set_ylabel('current (nA)',style='italic')
        ax2 = fig.add_subplot(212) # 29
        ax2.plot(self.timeseries,self.V,c='k') # 30
        ax2.set_title('integrate and fire voltage
response',style='italic') # 31
        ax2.set_xlabel('time (ms)',style='italic') # 32
        ax2.set_ylabel('voltage (mV)',style='italic') # 33
        plt.tight_layout() # 34
        #plt.show() # 35
        plt.savefig('Integrate and fire voltage response.png')
# 36
```

```matlab
classdef neuron %2
    properties %97
            C = 0.281; %4
            gL = 0.030; %5
            vR = -60.6; %6
            vT = -50.4; %7
            currentInj; %99
            T; %99
            timeseries; %99
            V; %99
    end %97
    methods %97
        function self = neuron %98
            self; %98
        end %98
        function self = create_injection_current(self) %8
            self.currentInj = cat(2,zeros(1,10),[[1:100]./100]); %9
            self.T = length(self.currentInj); %10
        end %97
        function self = leaky_integrate_and_fire(self) %11
            self.timeseries = linspace(1,self.T, self.T); %12
            self.V = ones(1,self.T).*self.vR %13
            ii = 1; %14
            while ii < (max(self.T)-2) %15
            dV = (-self.gL.*(self.V(ii) - self.vR) + ...
            self.currentInj(ii))./self.C; %16
            self.V(ii+1) = self.V(ii)+dV; %17
            if self.V(ii+1) >= self.vT %18
                self.V(ii+1) = 20; %19
                self.V(ii+2) = self.vR; %20
            ii = ii + 1; %21
            end %18
            ii = ii + 1; %22
            end
        end
        function plot_neuron(self) %23
            figure %24
            ax = subplot(2,1,1) %25
            plot(self.timeseries,self.currentInj,'color','k') %26
            title('current injection','FontAngle','italic') %27
            ylabel('current (nA)') %28
            ax2 = subplot(2,1,2) %29
            plot(self.timeseries,self.V,'color','k') %30
            title('integrate and fire voltage response', 'FontAngle',...
            'italic') %31
            xlabel('time (ms)') %32
            ylabel('voltage (mV') %33
            %shg %35
print('-dpng','-r300','Integrate and fire voltage response'); %36
            end
        end
    end
```

Now that we have declared the class Neuron, we instantiate it and create an object (i.e., we create one object of the type Neuron). We call it - myFirstNeuron - you can call this from a script or the command line [1]. All methods and objects in the *self* object can now be called by using the variable myFirstNeuron. For example, myFirstNeuron.vR will print -60.6.

We then call the Neuron method create_injection_current, so that now the variables currentInj and T are associated with myFirstNeuron. You'll notice that in calling these methods, we do not have to specify self as an input. POM works in a way such that the first input in the class declaration is self, but when you call these methods later, they do not require you to input them again. We then call the Neuron method leaky_integrate_and_fire, where the variable **self** above is a stand-in for the myFirstNeuron. Calling this method will result in a newly updated myFirstNeuron.V, which is what we care about. This is plotted by calling myFirstNeuron.plot_neuron. The output of this method is seen in Fig. 6.3.

Pseudocode	1. Create an instance of the class Neuron, calling it "myFirstNeuron" 2. Create the current to "inject" into the neuron 3. Call the *Neuron* method leaky_integrate_and_fire 4. Call the *Neuron* method plot_neuron
Python	```myFirstNeuron = Neuron() # 1 myFirstNeuron.create_injection_current() # 2 myFirstNeuron.leaky_integrate_and_fire() # 3 myFirstNeuron.plot_neuron() # 4```
MATLAB	```myFirstNeuron = neuron() %1 myFirstNeuron = myFirstNeuron.create_injection_current %2 myFirstNeuron = myFirstNeuron.leaky_integrate_and_fire %3 myFirstNeuron.plot_neuron %4```

The result of our first class Neuron is that we can observe a simple relationship between input (current) and output (voltage, spikes) of this neuron.

What if we are interested in modulating the magnitude of the current injected to see how this affects the number of evoked spikes? To do this, we create a new class Neuron2, which is similar to Neuron, but with a few key modifications: When creating the method create_injection_current we add an input parameter mag. In Python, we can say mag= 1, which means that if we do not specify an input later on when calling this method, it will use 1 as the default value of mag. Inside this method, we simply multiply our ramping stimulus by the value mag, to amplify the injected current. Inside the method leaky_integrate_and_fire, we add two lines of code: we initialize a spikeCounter at the start, and increment it every time we encounter a spike. In this way, we can count the number of spikes evoked for a particular input stimulus.

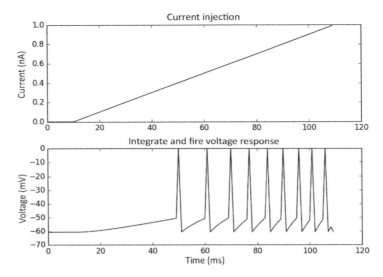

FIGURE 6.3 Simulated ramp injection current and spiking output. Top panel: Current injection. Bottom panel: Voltage response of neuron to the injected current. We can see that the neuron begins to spike more as the injected current magnitude increases.

Pseudocode

1. Import relevant packages
2. Declare the class called *Neuron2*
3. Initialize the class
4. Declare the variable for capacitance
5. Declare the variable for leak conductance
6. Declare the variable for spike threshold
7. Declare the variable for resting voltage
8. Create method create_injection_current with input variable mag
9. Create an array to simulate a current to inject, amplified by mag
10. Calculate the length of the injected current
11. Create method leaky_integrate_and_fire
12. Create the values for the timeseries
13. Set the voltage V equal to the resting voltage, vR
14. Initialize the index
15. Initialize the spikeCounter variable
16. For values less than the length of time minus 2
17. Calculate the change in voltage as a function of Equation 6.1,
18. Set the following index value to the present voltage plus dV
19. Check if spike threshold has been crossed
20. If so, set the voltage to 20
21. Reset the voltage to the resting potential
22. Increment the index in time
23. Add one to the variable spikeCounter
24. 24 Increment the index in time

Python	MATLAB

```python
import numpy as np # 1
import matplotlib.pyplot as plt # 1

class Neuron2(): # 2
    def __init__(self): # 3
        self.C= .281 # 4
        self.gL = .030 # 5
        self.vR = -60.6 # 6
        self.vT = -50.4 # 7

    def create_injection_current(self,mag=1): # 8
        self.currentInj = np.arange(100)/100.*mag # 9
        self.T = len(self.currentInj) # 10

    def leaky_integrate_and_fire(self): # 11
        self.timeseries = np.linspace(0,self.T-1,self.T) # 12
        self.V = np.ones(self.T)*self.vR # 13
        ii=0 # initial the index counter # 14
        self.spikeCounter=0 # 15
        while ii < self.T-2: # 16
            dV = (-self.gL*(self.V[ii] -
self.vR)+self.currentInj[ii])/self.C # 17
            self.V[ii+1]=self.V[ii]+dV # 18
            if self.V[ii+1]>self.vT: # 19
                self.V[ii+1]=20 # 20
                self.V[ii+2]=self.vR # 21
                ii+=1 # 22
                self.spikeCounter+=1 # 23
            ii+=1 # 24
```

```matlab
classdef neuron2 %2
    properties
        C = 0.281; %4
        gL = 0.030; %5
        vR = -60.6; %6
        vT = -50.4; %7
        currentInj; %99
        T; %99
        timeseries; %99
        V; %99
        spikeCounter = 0; %99
    end
    methods
        function self = neuron2;
            self;
        end
        function self = create_injection_current(self,mag) %8
            self.currentInj = ...
cat(2,zeros(1,10),[[1:100]./100].*mag); %9
            self.T = length(self.currentInj); %10
        end
        function self = leaky_integrate_and_fire(self) %11
            self.timeseries = linspace(1,110, 110); %12
            self.V = ones(1,self.T).*self.vR %13
            ii = 1; %14
            while ii < (self.T-2) %15
                dV = (-self.gL.*(self.V(ii) - self.vR) + ...
self.currentInj(ii))./self.C; %16
                self.V(ii+1) = self.V(ii)+dV; %17
                if self.V(ii+1) >= self.vT %18
                    self.V(ii+1) = 20; %19
                    self.V(ii+2) = self.vR; %20
                    ii = ii + 1;
                    self.spikeCounter = self.spikeCounter +1;
                end
                ii = ii + 1;
            end
        end
    end
end
```

To visualize the relationship between many different input magnitudes and the spiking output, we create a standalone function called `plotFI`, which takes as inputs `currentMags` and `spikes`. Each of these should be a one-dimensional array (vector). The function creates a scatter plot of the number of spikes evoked for each magnitude of injected current, using plotting methods that we are already familiar with.

Pseudocode

---

25. Create method plotFI outside of the object with inputs currentMags, spikes
26. Create the figure
27. Add subplot
28. Plot scatter points of spikes vs current magnitude
29. Set the x-label
30. Set the y-label
31. Set the title
32. Show the figure (uncomment line to do this)
33. Save the figure

Python	MATLAB

```python
def plotFI(currentMags,spikes): # 25
    fig=plt.figure() # 26
    ax = fig.add_subplot(111) # 27
    ax.scatter(currentMags,spikes,c='k',edgecolor='w',s=50) # 28
    ax.set_xlabel('current injection maximum
(nA)',style='italic') #29
    ax.set_ylabel('number of spikes',style='italic') # 30
    ax.set_title('Firing as function of current
injected',style='italic') # 31
    #plt.show() # 32
    plt.savefig('Firing as function of current injected.png') #
33
```

```matlab
function [] = plotFI(currentMags,spikes)
    figure
    h = ...
plot(currentMags,spikes,'o','MarkerFacecolor','k', ...
'Markeredgecolor','w','markersize',10);
    xlabel('current injection maximum (nA)')
    ylabel('number of spikes')
    title('Firing as function of injected current')
    shg
    print('-dpng','-r300','Firing as function of ...
injected current'); %36
end
```

To simulate how the neuron responds to many different stimuli, we iterate over many injected magnitudes (in Python, we make use of the `numpy` function `arange`, creating an array of values between 0.1 and 10, with step increment 0.1). Iterating over all of these magnitudes in a for loop, we create an object `mySecondNeuron` (which is an instance of class `Neuron2`), calling its methods for each value `mag`, and for each value `mag` we append the number of spikes evoked (`spikeCounter`) to our list (vector) spikes. After looping through each value in `currentMags`, we visualize the relationship using the function `plotFI`, defined above. The output of this is seen in Fig. 6.4.

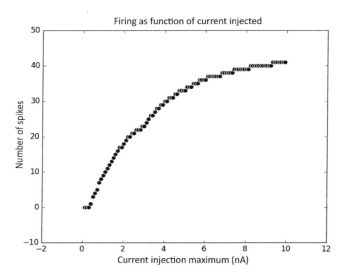

FIGURE 6.4   Simulating evoked firing rate as a function of increasing the maximum current injected.

Pseudocode   1. Initialize list that will hold number of spikes for each stimulus
              2. Create array of injection current magnitudes
              3. For each injection current magnitude
              4. Create an instance of the class Neuron
              5. Create the current to "inject" into the neuron
              6. Call the *Neuron* method leaky_integrate_and_fire
              7. Append the spikeCounter value of the class instance to the spikes list
              8. Plot the firing (F) vs the injected current (I)

Python
```
spikes=[] # 1
currentMags=np.arange(0.1,10,0.1) # 2
for mag in currentMags: # 3
    mySecondNeuron = Neuron2() # 4
    mySecondNeuron.create_injection_current(mag) # 5
    mySecondNeuron.leaky_integrate_and_fire() # 6
    spikes.append(mySecondNeuron.spikeCounter) # 7
plotFI(currentMags,spikes) # 8
```

(Continued)

```
MATLAB          spikes=[] %1
                currentMags= 0.1:0.1:10; %2
                for mag = currentMags %3
                    mySecondNeuron = neuron2(); %4
                    mySecondNeuron = mySecondNeuron.create_injection_current(mag); %5
                    mySecondNeuron = mySecondNeuron.leaky_integrate_and_fire(); %6
                    spikes = cat(2,spikes,mySecondNeuron.spikeCounter); %7

                end
                plotFI(currentMags,spikes) %8
```

Now that we created functional neuron classes and know how to measure their behavior, play around with the parameters a bit. What happens if you modify capacitance or conductance? What happens if you change the current you inject? Our model makes clear predictions that could be tested experimentally, with real neurons. If the empirical results are in line with the predictions, the model gains support—it seems to capture something relevant about the behavior of the system. If the model fails to predict the empirical results, or if the empirical results are surprising in light of the model predictions, the model will have to be modified. Data have primacy. If data and model disagree, it is usually the fault of the model, not the data (if the data were properly logged).

MODELING THE RESPONSE PROPERTIES OF THIS NEURON

Modeling the neuron's response properties is a useful approach for approximating how the relation between the input to the neuron and its spike output can be described, hopefully by just a few numbers. If for every possible input to a neuron, we had to calculate the output iteratively each time, this might become computationally cumbersome (especially in the extreme case of a large network of neurons). Thus we create a model such that the spike output can be predicted even more simply than by the five parameters we used to create our biophysical model. Models are only useful when the number of parameters we use in an equation is less than the number of parameters that govern the behavior of the system of interest (or else we fall victim to overfitting) and are only practical if computationally cheap.

We first consider a three-parameter exponential function to account for the total number of spikes Y evoked as a function of x. The parameter A corresponds to the maximum number of evoked spikes, built around the idea that neurons have some maximum biophysical limit to the number of spikes that can be evoked due to intrinsic properties such as the refractory period, whereas the parameters B and C correspond to how quickly the spike maximum is reached for each value x. For example, for large values of C, the term e^{-C*x} approaches zero quickly, so the spike output Y is equal to the maximum number of spikes A for any current input. We can think of these parameters B and C as controlling the *gain* of the neuron.

$$Y = A - B * e^{-c*x} \tag{6.2}$$

To fit the model in Eq. (6.2), we write three functions in POM: one function is the equation itself, one calculates the residuals (or, the total error in the model), and one actually fits parameters using a least squares method. Note that in MATLAB, it is easier to write these functions as separate files, although it *could* be done inline. --- in the code below indicates a separate function file. Also, we don't need the function that calculates the residuals, as MATLAB's lsqcurvefit function calculates residuals implicitly already.

Pseudocode

---

1. Import package for least squares fitting
2. Create function pevalEq that requires x to be an array and p to be a list containing 3 values (parameters A, c, B).
3. Returns Y from Eq 6.2.
4. Create function residualsEq, a function that calculates the residuals between any data set y and the output of the model given parameters p
5. Calculate the error between y and what is yielded by the function, given the coefficients in p
6. Return the error
7. Create a function to perform a least squares evaluation in which the sum of the squared deviations between data and model prediction is minimized
8. Use least squares method for finding good coefficients for p
9. Return the function with optimal coefficients

Python	MATLAB

```python
from scipy.optimize import leastsq # 1

def pevalEq(x,p): # 2
    return p[0] - p[2]*np.exp(-p[1]*x) # 3

def residualsEq(p,y,x): # 4
    err = (y - (p[0] - p[2]*np.exp(-p[1]*x))) # 5
    return err # 6

def eqFit(x,p0,y_meas): # 7
    plsq = leastsq(residualsEq,p0,args=(y_meas,x))#8
    return pevalEq(x, plsq[0]) # 9
```

```matlab
function [y] = pevalEq(p,x)
y = p(1)-p(3).*exp(-p(2).*x);
end
---
function [fittedCurve] = eqFit(x,p0,y_meas)
plsq = lsqcurvefit(@pevalEq,p0,x,y_meas);
fittedCurve = pevalEq(plsq,x);
end
```

We now evaluate the simulated neural spike patterns as a function of input current and fit the output response with both the exponential function of Eq. (6.1), and the general *polynomial* of Eq. (6.2), using a few different orders. The order of a polynomial is also called its "degree" and determined by the highest power or exponent involved:

$$Y = A + B * x + C * x^2 + D * x^3 + E * x^4 + \ldots \tag{6.3}$$

where the variables A, B, C, D, and E are coefficients of a general polynomial, and the ellipsis indicates that we can keep adding coefficients (F, G, and so on) to higher exponents of the input, as many as we want.

Up until this point, we have taken data at face value—the underlying idea being that data speak for themselves, and are effectively unimpeachable. We are now starting curve fitting (trying to account for a dataset—which is almost always noisy and messy) with a function (which is usually simple and pristine). Science has three fundamental goals: describing, explaining, and predicting phenomena in the natural world. Describing data is the job of statistics. Predicting will be covered in the Chapter 7, Regression. Curve fitting falls under the rubric of explaining the world. We are trying to transcend the data and identify the processes that generated it. If we have done so, we have understood something about the natural world. Like so many other things in this book, this is an exercise in platonism. We use noisy data from the messy world to transcend it, by uncovering the functions that underlie them. These functions live in the world of ideas. The underlying philosophy of modeling hardliners is that data can't be taken at face value as they are always corrupted by noise and who knows what else—they are simply used to constrain the model, but it is the model that really matters. More on this in the thought at the end of this chapter, but for now, we note that we are playing with fire—the "data" we use for our curve fitting exercise were generated by our simulated neurons. Have we already lost touch with reality altogether, before we even really started? It can happen quickly. We are doing this for didactic reasons, because data generated by simulations are almost always cleaner, but it is important to keep in mind that it is not actually real.

The point of curve fitting is to find the parameters that yield the closest fit between model and data. Data are simply used to find the right parameters to put into the model.

When *fitting* a model to data, we need some way (besides just visual inspection) to calculate how well the model describes the data.

There are several metrics in popular use to determine the goodness of fit between data and a model. One of these is the *root mean squared error* (RMSE), simply the mean of the difference between the actual datapoints and the points generated by the model, first squared (to get rid of sign), then square-rooted (to get rid of the effects introduced by squaring on magnitude). Let's calculate the root mean squared error by fitting polynomials of increasing order to a sine wave (Fig. 6.5).

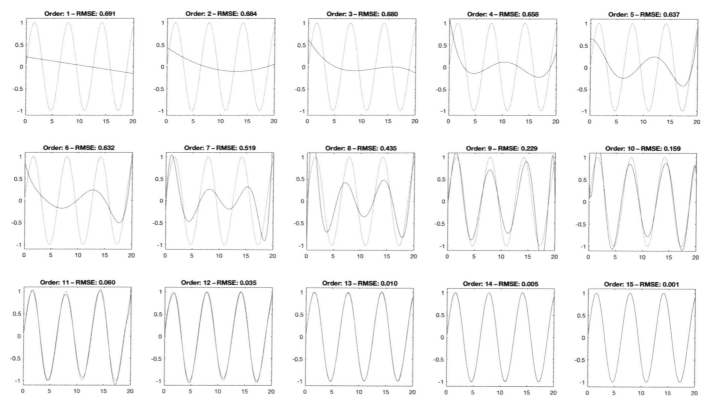

FIGURE 6.5   Fitting a polynomial to a sine wave.

1. Define an x-base
2. Take the sine wave of that
3. Decide the highest order polynomial we want to create
4. Initialize a variable that will contain root mean square errors of the right size
5. Open a new figure
6. Open a new subplot
7. Fit a polynomial to the data, of order nn
8. Initialize the model output
9. Recursively compute the model output, using the general polynomial formula
10. Loop through all polynomials
11. Plot the sine function as a function of x, in blue
12. Turn the hold on so we can plot on top of that
13. Plot the model output (the result of the curve fitting, for a given polynomial), in red
14. Specify limits of y-axis to plot and show graph
15. Calculate the root mean square error and capture it in the container variable RMSE
16. Put a title on the subplot, indicating order of the polynomial and RMSE
17. Loop through all orders of polynomial
18. Open a new figure
19. Plot RMSE vs. order number as a thick, black line
20. Add axis labels and increase the font size

Python	MATLAB

```python
x = np.arange(0,20,0.1)#1
y = np.sin(x) #2
highestOrder = 15 #3
RMSE = np.zeros(highestOrder)#4
f = plt.figure(figsize=(10,10)) #5

for nn in range(highestOrder):#17
    ax = plt.subplot(3,5,nn+1) #6
    p = np.polyfit(x,y,nn+1) #7
    y2 = np.zeros(len(x)) #8
    for ii in range(nn+2): #10
        y2 = y2 + p[ii]*x**(nn-ii+1) #9
    ax.plot(x,y) #11
    ax.plot(x,y2,c='r') #13
    ax.set_ylim([-1.1, 1.1]) #14
    RMSE[nn] = np.sqrt(np.mean((y-y2)**2)); #15
    ax.set_title('Order: '+ str(nn)+ ' - RMSE: '+
str(RMSE[nn])[:5],fontsize=8) #16

f=plt.figure()
plt.plot(range(highestOrder),RMSE,c='k',lw=3) #19
plt.xlabel('Polyromial order',fontsize=20) #20
plt.ylabel('RMSE',fontsize=20) #20
```

```matlab
x = 0:0.1:20; %1
y = sin(x); %2
highestOrder = 15; %3
RMSE = zeros(highestOrder,1); %4
figure %5

for nn = 1:highestOrder %17
    subplot(3,5,nn) %6
    p = polyfit(x,y,nn); %7
    y2 = 0; %8
    for ii = 1:nn+1 %10
        y2 = y2 + p(ii).*x.^(nn-ii+1); %9
    end %10
    plot(x,y) %11
    hold on %12
    plot(x,y2,'color','r') %13
    ylim([-1.1 1.1]) %14
    shg %14
    RMSE(nn,1) = sqrt(mean((y-y2).^2)); %15
    title(['Order: ', num2str(nn), ' - RMSE: ', ...
    num2str(RMSE(nn,1),'%1.3f')]) %16
end %17

figure %18
plot(1:highestOrder,RMSE,'color','k','linewidth',3) %19
xlabel('Polynomial order') %20
ylabel('RMSE') %20
set(gca,'FontSize',20) %20
```

If you add enough terms, polynomials can approximate any function. Note that with polynomials of order 15, we can reduce RMSE between the polynomial and the sine wave to effectively zero (Fig. 6.6).

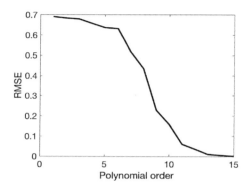

FIGURE 6.6   Root mean squared error as a function of polynomial order.

But also note that the ever-declining RMSE leads us down the garden path here. The fit gets better and better, but are we understanding the system better and better? A low RMSE by itself does not necessarily mean that you understand the system with your polynomial model. In this case, the functional relationship in the data is sinusoidal. We know because we created it. But the polynomial model completely misses that and mischaracterizes the functional relationship as polynomial in nature. Why does this matter? The 15th order polynomial seems to produce a fit that is close to perfect. What's not to like? The problem is that we built a model that perfectly fits *these* datapoints, not the actual sine function. In other words, it won't be robust to any other values other than the ones we used to fit the polynomial. Say we reuse the p-values from the 15th order polynomial, but plot the sine wave just a little farther:

Pseudocode

```
1. x base just extended by a tiny bit, by 10%
2. Take sine of x, like before
3. Determine the product of the polynomial, reusing p and nn from before
4. Open a new figure
5. Plot the sine wave in blue
6. Hold on, so we can plot other things on top of it
7. Plot the output of the polynomial over the same x base
8. Calculate root mean square error
9. Format the plot to show the x-base
```

Python	MATLAB
| ```python
x = np.arange(0,22,.1) #1
y = np.sin(x) #2
y2 = np.zeros(len(x)) #8
for ii in range(nn+2): #10
    y2 = y2 + p[ii]*x**(nn-ii+1)

f=plt.figure() #4
plt.plot(x,y,lw=2) #5
plt.plot(x,y2,c='r',lw=2) #7
rmse = np.sqrt(np.mean((y-y2)**2)) #8
plt.xlim([min(x), max(x)]) #9
``` | ```matlab
x = 0:0.1:22 %1
y = sin(x); %2
y2 = 0; %3
for ii = 1:nn+1 %3
    y2 = y2 + p(ii).*x.^(nn-ii+1); %3
end %3

figure %4
    plot(x,y,'linewidth',2) %5
hold on %6
plot(x,y2,'color','r','linewidth',2) %7
rmse = sqrt(mean((y-y2).^2)) %8
xlim([min(x) max(x)]) %9
``` |

We even had to give up our near perfect RMSE, which is now 3.27, unacceptably high (Fig. 6.7).

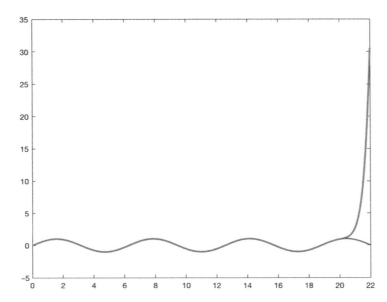

FIGURE 6.7   The perils of fitting higher-order polynomials, illustrated. The polynomial fit those exact points, not the sine function. Extending the sine function just a little bit revealed that our perfect fit was a comforting illusion. Things can go off the rails quickly.

Here, we fit to a pure sine wave, without any noise. If we had added some noise, our polynomial would still be able to account for *these points* perfectly, but not generalize to other situations where the noise won't be exactly the same. In that sense, noise is salutary, as it will allow us to build more robust models, models that generalize (Taleb, 2012).

In addition, note that models are more credible if they can achieve low RMSE with less terms—you should not be surprised that your higher-ordered model can account for any data almost perfectly—as von Neuman observed: "With four parameters I can fit an elephant, and with five I can make him wiggle his trunk" (Dyson, 2004).

Another common metric for determining how well the model accounts for the data is r^2. We'll be talking much about this in the next chapter on regression, for now, an r^2 value of 0 indicates that the model accounts for none of the variation in the data and a value of 1 indicates that it accounts for all of the variation in the data. To get our value of r^2, we introduce the `function rsquared`. This makes use of the scipy methods "stats," and returns the r^2 value of a linear regression between two curves (again, we cover this in more depth in chapter: Regression). Models with more terms will always account for more variation in the data than models with less terms. How to correct for this is subject to intense debate, but the Akaike Information Criterion (AIC) is perhaps used most commonly (Akaike, 1974, 2011).

Again, we play our dangerous game—simulating the behavior of a neuron as an instantiation of the class `neuron2` we declared above, but now we add some additional functions such as `rsquared`. We update `plotfit`. We do assume that the rest of the code in this chapter was already run (Fig. 6.8).

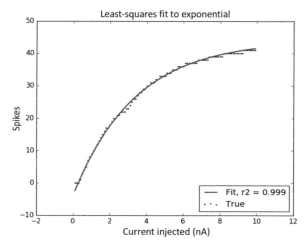

FIGURE 6.8   Least squares fit to firing-current (F-I) response of neuron. We see that the firing response curve for this particular neuron seems well described by an exponential function.

Pseudocode

0. Import Python stats package
1. Declare function rsquared
2. Use stats function to perform linear regression between input variables x and y
3. Return r_val, squared
4. Declare function plotFit
5. Create figure
6. Copy y to y_true
7. Add noise to y_true to get y_noise_added
8. Guess initial values
9. Predict y given x, initial guesses of p, and a noisy y
10. Plot a line prediction of y that fits Equation 6.2
11. Scatterplot of x and the true values of y
12. Add a title
13. Add a legend
14. Add xlabel
15. Add ylabel
16. Show the figure
17. Save figure

Python

```python
from scipy import stats #0
def rsquared(x,y): #1
    Slope,int_,r_val,p_val,std_err = stats.
linregress(x,y)#2
    return r_val**2 # 3

def plotFit(x,y): # 4
    fig = plt.figure() # 5
    y_true = y # 6
    y_noise_added = y_true + np.random.randn(len(x)) #7
    p0 = [40,.4,40] # 8
    y_predict = eqFit(x,p0,y_noise_added) #9
    plt.plot(x, y_predict,c='b',lw=2) #10
    plt.scatter(x,y_true,c='k',s=2) #11
    plt.title('Least-squares fit to exponential') #12
    plt.legend(['Fit, r2 = '+ str(np.round(rsquared
(y_predict, y_true),3)),'True'], loc='lower right') #13
    plt.xlabel('current injected (nA)',style='italic') #14
    plt.ylabel('spikes',style='italic')#15
    plt.show() #16
    plt.savefig('Least squares fit to exponential.png') #17
```

MATLAB

```matlab
function [RS] = rSquared(x,y) %1
rs = corrcoef(x,y); %2
RS = rs(1,2).^2; %3
end %1
-----
function [] = plotFit(x,y) %4
    fig = figure; % 5
    y_true = y; % 6
    y_noise_added = y_true + randn(1,length(x)) %7
    p0 = [40,.4,40]; % 8
    y_predict = eqFit(x,p0,y_noise_added); %9
    h1 = plot(x, y_predict, 'color','b', 'linewidth',2); %10
    hold on
    h2 = plot(x,y_true,'.','color','k','markersize',20); %11
    title('Least-squares fit to exponential') %12
    legend([h1,h2],['Fit, r^2 = ', ...
    num2str(rSquared(y_predict,y_true))],...
    'Ground truth','Location','SouthEast') %13
    xlabel('current injected (nA)') %14
    ylabel('spikes') %15
    shg %16
    print('-dpng','-r300','Least squares fit to ...
    exponential'); %17
end %4
```

Pseudocode	1. Initialize list spikes
	2. Create list of injection current magnitudes
	3. For each magnitude in list of injection currents
	4. Create an instance of the class Neuron, calling it "myFirstNeuron"
	5. Create the current to "inject" into the neuron
	6. Call the *Neuron* method leaky_integrate_and_fire
	7. Append the number of spikes evoked to the list spikes
	8. Run method plotFit

Python
```
spikes=[] # 1
currentMags=np.arange(0.1,10,0.1) # 2
for mag in currentMags: # 3
    classInstance = Neuron2() #4
    classInstance.create_injection_current(mag) #5
    classInstance.leaky_integrate_and_fire() #6
    spikes.append(classInstance.spikeCounter) #7
plotFit(currentMags,spikes) #8
```

MATLAB
```
spikes=[] %1
currentMags= 0.1:0.1:10; %2
for mag = currentMags %3
    mySecondNeuron = neuron2(); %4
    mySecondNeuron = mySecondNeuron.create_injection_current(mag); %5
    mySecondNeuron = mySecondNeuron.leaky_integrate_and_fire(); %6
    spikes = cat(2,spikes,mySecondNeuron.spikeCounter); %7
end
plotFit(currentMags,spikes) %8
```

This is as far as we want to go in terms of OOP to model biophysical properties of single neurons by simulation. We think it is fair to say that we reached "1" in that regard. What lies beyond 1? Lots of things—in all directions. In terms of biophysics, all of these processes are understood on a much deeper level. For instance, we could have included a more sophisticated model for the membrane, including all individual ion channels and their conductances (there is a reason the resting potential is close to the reversal potential of potassium… (Doyle et al., 1998)). The generation of action potentials via ion flux is understood in much more detail (Hodkin & Huxley, 1952). We haven't even mentioned the cable equation—a computational neuroscience classic, which allows to understand current flow in dendrites in electrotonic space (Rall, 1969). Similarly, we don't come close to fully use the power of OOP here, for instance we didn't do anything with inheritance. *Inheritance* is where the real power of OOP lies as it allows for the efficient creation of related classes without having to specify everything again. Finally, modeling, simulations, and curve fitting are entire fields unto themselves.

⌘. *Pensee on the primacy of data:* There is nothing wrong with modeling and simulation as long as you never forget what you are doing and that reality is the ultimate arbiter of whether simulations and models are correct or even useful. It is easy for the theoretical neuroscientist to get lost in a *luftschloss* of highly abstract models that are validated with "simulated data," not tested with actual data. Many theoretical problems are best solved by dropping unrealistic assumptions about how the system works—assumptions one wouldn't have made in the first place if one was in close touch with the data. Thus, we think that a deeper appreciation of the role of data will be salutary in this regard. Neuroscience is far from alone in this predicament: Assuming an ideal economic market, it can be shown that the ideal market indeed works ideally (Ricardo, 1891). Economists (enthralled by the sheer mathematical elegance of their models) might be forgiven for confusing them with reality after a while. However, the real-life consequences of such behavior can be devastating. For instance, modeling risk with normal distributions—unsuitable for much of social science, a fact underappreciated by some quants who entered the field from physics—seriously underestimates tail risk, resulting in devastating consequences for the global economy in 2008 (Taleb, 2007). It is critical not to confuse models with data. Just because your model "works" and is elegant doesn't mean that it captures anything that corresponds to the real world. Models have many advantages, they are usually simple and built from known principles and they complement data nicely, but they are no substitute for data. In a similar vein, modeling choices do matter. In most biophysical models, neural membranes are conceptualized as a battery, axons as wires. That's fine, but if that is all that is required to create a mind, we have to be ready to accept that your remote control has a mind of its own. In models, always be suspicious of what is represented and what is *not* represented (and how much that might matter). It is easy to get lost in the beauty of one's ideas and the formalism needed to express them. Data, recorded well, provide a necessary corrective (Brzychczy & Poznanski, 2013). Overall, epistemic modesty and humility are paramount. At this point, we cannot even explain how single neurons work. Try to avoid the temptation of trying to explain how consciousness works, no matter how many books claiming to do so might sell (Dennett, 1993; Pinker, 1999). Above all, don't underestimate the challenge of understanding systems that are inherently complex and thus better able to resist analytical approaches than we would like to admit (Fig. 6.9).

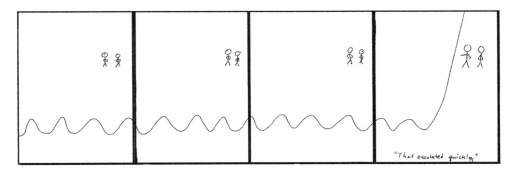

FIGURE 6.9 Overfitting.

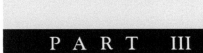

PART III

GOING BEYOND THE DATA

CHAPTER

7

Regression

Science is concerned with the description, explanation, and prediction of natural phenomena. We have covered description and explanation in previous chapters. Here, we will focus on prediction. In particular, we want to make predictions in a principled (and hopefully accurate) fashion about things we have not observed yet on the basis of things that we have.

The technical term for this is called "regression" (Galton, 1886), for reasons that will become clear shortly.

Some data scientists believe that statistics describes the past but that machine learning predicts the future. But this is not quite accurate. The principled prediction of the future has a long history that far precedes machine learning and, until very recently, was wholly subsumed under the rubric "statistics."

Interestingly, the original use case of regression analysis addressed, and solved, a potentially fatal problem in Darwin's Theory of Evolution. Charles Darwin believed in the primacy of data. He stated that he had "no faith in anything short of actual measurement and the Rule of Three." What is meant by this rule of three is a simple arithmetic rule (popular on standardized tests) such as if a is to b as c is to x, what is x? The potential problem with the Theory of Evolution is that if this were empirically true, runaway variation would wreck his theory: Tall people would give birth to tall children whereas short people would give birth to short ones. Over time, variability would escalate. Galton showed that this is not the case, that variability is largely stable across generations and also why, namely due to "regression". Regression means that if measures are not perfectly correlated, the rule of three actually doesn't hold (there is a regression to the mean, more on this later), ironically saving Darwin's theory of evolution—if it held, it would be a victim of runaway cumulative overdispersion (Stigler, 2016).

The figure of merit in regression is usually r squared or r^2. It denotes the proportion of variation in the variable that is to be predicted that can be accounted for by the model. In other words, if r^2 is 1, we can perfectly predict all values in the data (a situation that should always make you suspicious of overfitting, as there will always be (measurement) noise, which would

Neural Data Science.
DOI: http://dx.doi.org/10.1016/B978-0-12-804043-0.00007-6

make it less likely that the model generalizes). If r^2 is zero, our model fails to predict any of the data. Regression generally comes down to fitting a function that represents a model. In this chapter, we will fit lines to implement linear regression and sigmoidal curves for logistic regression. Both of these are popular regression types and serve as a good starting point to get familiar with the topic. Without further ado, we start with linear regression to predict phenomena in auditory cortex.

DESCRIBING THE RELATION BETWEEN SYNAPTIC POTENTIALS AND SPIKES

In Chapter 3, Wrangling Spikes Trains, we analyzed spikes, and in Chapter 4, Correlating Spike Trains, we analyzed units across many electrodes. In Chapter 5, Analog Signals, we looked at ways to handle analog signals, and in Chapter 6, Biophysical Modeling, we introduced the relation between the intrinsic cell properties and the spike output. We build on many of the concepts introduced before to now closely examine how the shape of synaptic inputs is related to the shape of the spike output.

In this chapter, we dive into real single cell data: neurons recorded in whole-cell configuration in vivo from the auditory cortex (see Fig. 6.2 for visualization of whole cell recording). The intracellular voltage of neurons is recorded at their resting potential, while tones with a duration of 50 ms are presented. The resulting voltage deflections and spike outputs are considered auditory-evoked responses. In this dataset, a range of *carrier frequencies* and *tone intensities* was used, but we aren't interested in, say, the neuron's frequency tuning (frequency tuning in auditory systems is a way of describing how a particular neuron responds to different frequencies of sounds). We are interested in a more general question: what is the relationship between the subthreshold potentials and the spiking output? One useful way to determine how related two things are to perform a regression analysis.

In the dataset, you will find *subthreshold* (meaning *nonspiking*) postsynaptic potentials (PSPs) as well as spike times. Note that the trials in which the spike occurred are noted simply as the time of the spike, and do not include the voltage trace. So these are essentially two separate datasets from the same neurons: the traces in which PSPs occurred (and spikes did not), and the spike times for trials where spikes occurred. Thus, there are a different number of PSP voltage traces than spike times.

A regression analysis between the PSPs and the spike outputs (represented by the poststimulus time histogram, PSTH) will tell us something about how related these two properties are. More intuitively, the regression analysis will tell us how similar the shape of the PSP is to the shape of the PSTH. If the shape of the PSP is very similar to the shape of the PSTH, then the PSP would be a good predictor of the PSTH. That is, if the shapes of the PSP and PSTH are similar, then we would be able to use measurements of the synaptic inputs to a neuron to predict the spiking output.

For this analysis, we begin by loading the file `Psptospikedict.pickle` from the companion site. As the name suggests, this is a dictionary that contains a few fields. To view these fields, we look at the keys. In MATLAB (to avoid getting into a pickle) we load the workspace `pspsAndSpikes.mat`.

Python	MATLAB
``import numpy as np
import matplotlib.pyplot as plt
import pickle
with open('Psptospikedict.pickle', 'rb') as handle:
pspsAndSpikes = pickle.load(handle)
pspsAndSpikes.keys()```	```load('pspsAndSpikes.mat')```

We see that in the Python dictionary pspsAndSpikes there are three keys: ['pspTime', 'pspTraces', 'spikeTimesByTrial']. In MATLAB, we observe that the variable pspTime contains the timebase, the variable pspTraces contains 100 voltage traces on that timebase, and spikeTimesByTrial is a cell array containing the time when spikes occurred during individual trials.

Let's look at the PSTH of those spikes, like we did in Chapter 3, Wrangling Spikes Trains and Chapter 4, Correlating Spike Trains (Fig. 7.1).

Pseudocode

```
0. Unpack the spikes
1. Define an edges vector that will be used both to parse the spiketrain and serve as a time base later
2. Preallocate a spike catcher variable
3. Find spike times from a given trial and add them to the spike catcher variable
4. Do this for all trials
5. Parse the spike catcher variable by the time base
6. Open a new figure
7. Make a bar graph of the PSTH over the time base
8. Label the x axis
9. Set the title
```

Python	MATLAB
``spikes=pspsAndSpikes['spikeTimesByTrial'] #0
spikesUnpacked = [a for b in spikes for a in b]
binrange = np.arange(-30,100,1) #1
psth,bins,patches=plt.hist(spikesUnpacked,bins=binrange
,facecolor='k') #5

plt.xlabel('time (msec)');plt.ylabel('spike count') #8
plt.title('Peristimulus time histogram') #9``` | ```binRange = -30:100; %1
spikeCatcher = []; %2
edges = min(binRange):max(binRange);
for ii = 1:length(spikeTimesByTrial) %4
    spikeCatcher = cat(2,spikeCatcher,…
    spikeTimesByTrial{ii}); %3
end %4
PSTH = histc(spikeCatcher,binRange); %5
figure %6
bar(edges,PSTH)%7
xlabel('time (msec)'); ylabel('spike count') %8
title('Peri-stimulus time histogram') %9``` |

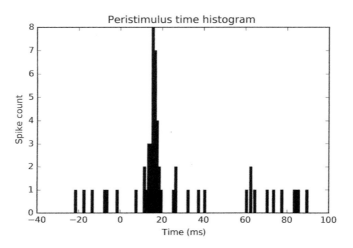

FIGURE 7.1   Here we do some good old-fashioned data wrangling.

The PSPs are labeled in the dictionary as `pspTraces`. We are interested in how the shape of this subthreshold potential relates to the shape of the PSTH. These are all in the timeseries of voltages from the neuron in which a spike did not occur.

We calculate the average PSP (`meanPsp`), then downsample it so that it has the same time resolution as the PSTHs (this is important for regression analysis, the two lines we compare have to be of the same length), to get the variable `downSampledPsp`. In the data file, natively, the PSTH has bin size 1 ms and the PSPs 0.1 ms. This downsampling is done by taking every 10th value, and averaging it with the nine values after each sampled value. This creates an output that is 10 times smaller than the input. In a way, it is like zooming out on the line, or the opposite of telling the computer to "enhance."

We use the histogram output variables `psth`, `binrange` from above to get an array `evokedPSTH` that represents just the evoked PSTH.

We also calculate the derivative of the PSP (`dPSP`) by using the function `diff`, which returns the point-wise difference of each value in the input array.

Because time on the x-axis ranges from –30 to 100 ms, we then select only the region of evoked response—i.e., the responses that occur between 0 and 50 ms (which is the time during which the stimulus was presented).

Pseudocode

1. Calculate the mean psp trace
2. Downsample the Psp by blending 10 samples together
3. Take the point to point differential of the mean downsampled PSP
4. Find index that corresponds to time = 0
5. Find index that corresponds to time = 50
6. Make new vector that corresponds to the snippet of the respective vectors that is in the range of interest
7. New timebase, from 0 to 50

Python	MATLAB
| ```
meanPsp = np.mean(pspsAndSpikes['pspTraces'],0) #1


downSampledPsp = [np.mean(meanPsp[ind_*10:ind_*10+10])
for ind_ in range(len(meanPsp)/10)] #2
dPSP = np.diff(downSampledPsp)#3

zeroIndex=np.where(binrange==0)[0][0]#4
fiftyIndex=np.where(binrange==50)[0][0]#5

evokedPSTH = psth[zeroIndex:fiftyIndex] #6
evokedPSP = downSampledPsp[zeroIndex:fiftyIndex]#6
evokedDPSP = dPSP[zeroIndex:fiftyIndex]#6
evokedRange = range(50) #7
``` | ```
meanPsp = mean(pspTraces); %1
indices = (0:130).*10+1; %2
for ii = 1:length(indices)-1 %2
downSampledPsp(ii) = mean(meanPsp(indices(ii):…
indices(ii+1)-1)); %2
end %2

dPSP = diff(downSampledPsp); %3
zeroIndex = find(binRange==0); %4
fiftyIndex = find(binRange==50); %5

evokedPSTH = PSTH(zeroIndex:fiftyIndex); %6
evokedPSP = downSampledPsp(zeroIndex:fiftyIndex); %6
evokedDPSP = dPSP(zeroIndex:fiftyIndex); %6
evokedRange = 0:50; %7
``` |

We now replot the evokedPSTH as a black line so that we can overlay other plots for comparative purposes. As we are interested in the relation between the synaptic inputs and the spike outputs, we plot the average subthreshold potential as a blue line. Back to our original question: what is the relationship between the shape of a synaptic input and its output (spiking) probability? We first analyze this graphically, by overlaying the average PSP and the PSTH on the same plot, making a double-y-axis plot.

Python	MATLAB
| ```
f, axarr = plt.subplots(1,1,figsize=(6,5)) #1
line1=axarr.plot(evokedRange,evokedPSTH,c='k',label='P
STH',lw=2)#2
plt.ylabel('spike count')#3
plt.xlabel('time (msec)')#4
ax2 = axarr.twinx()#5
line2 = ax2.plot(evokedRange,evokedPSP,c='b',label='PS
P',lw=2)#6

lines = line1+line2#7
lineLabels = [l.get_label() for l in lines]#8
axarr.legend(lines, lineLabels, loc=0,fontsize=10)#9
plt.ylabel('voltage (mV)')#10
plt.title('PSTH and Mean PSP')#11
plt.savefig('Figure 7.2 - PSTH and mean PSP.
png',dpi=500)#12
``` | ```
f = figure; %1
[ax, h1, h2] = plotyy(evokedRange,evokedPSTH,…
evokedRange,evokedPSP); %2, %6, %7
ylabel('spike count') %3
xlabel('time (msec)') %4
title('PSTH and Mean PSP') %11
h1.Color = 'k'; h1.LineWidth = 2; %13
h2.Color = 'b'; h2.LineWidth = 2; %13
legend([h1, h2], 'PSTH', 'PSP'); %9
print('-dpng','-r300','PSTH and mean PSP'); %12
``` |

Pseudocode

1. Create figure
2. Plot PSTH
3. Set ylabel
4. Set xlabel
5. Double the y-axis to have 2nd axis on right
6. Plot evoked PSP on 2nd y-axis
7. Hold both lines
8. Get the line labels
9. Set the legend according to line labels
10. Set the ylabel
11. Set the title
12. Save figure
13. Set line properties

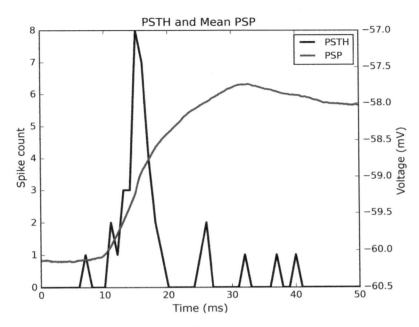

FIGURE 7.2 PSTH (in spikes) and mean PSP (in mV). Note the double y-axis!

At first glance, the blue line and the black do not appear to be very correlated. But looks can be deceiving. To test whether there is a statistical relationship between the PSP and the PSTH, we turn to a method called *linear regression*. In this case, linear regression will establish how many units of change in spiking activity line can be accounted for by a unit change in membrane voltage. In linear regression, we are interested in understanding how much *variance* in one variable can be described by *variance* in another (Fig. 7.2).

To do a linear regression, we use the Python package `scipy.stats.linregress` (MATLAB function `regress`). It returns several variables of interest: the slope, intercept, and *r*-squared value are what we are concerned with here. The slope and intercept are the variables m and b, respectively in the canonical equation of a line:

$$y = m * x + b \tag{7.1}$$

Hence *linear* regression. There are many kinds of regression, but simple linear regression finds the best fit line to a cloud of points where we are trying to predict one variable of interest from the known value of another variable.

Generally speaking, regression is about estimating parameters that govern the best fit of the geometric object to the data. These are beta-weights and in the case of a simple linear regression, they correspond to the intercept (β_0) and slope (β_1) of a line. Note that there is also an error term—there will be noise, here represented by epsilon (ε):

$$Y = \beta_0 + \beta_1 X_1 + \varepsilon \tag{7.2}$$

where X is the independent variable or predictor and Y is the dependent or predicted variable. β are the weights or parameters. β_0 governs the offset from the origin at $x = 0$ and β_1 the steepness of the slope.

Regression is driven by correlation. Say we want to estimate income on the basis of IQ and we know that the correlation between income and IQ is 0.42. This doesn't seem implausible—everything else being equal, someone smarter will probably make somewhat more money than someone less smart, but other things clearly play a role. In that case, the regression line will look like the red line in Fig. 7.3.

Blue dots represent individuals. The red line can be conceptualized as a conditional average—it minimizes the squared deviations between the dots and the line. The hashed black line simply connects vertices of standard deviations in each variable. Blue hashed lines delineate standard deviations in terms of IQ (vertical) and income (horizontal). If the correlation between the variables was 1, the red line would fall on the hashed black line, as the correlation is not perfect, a "regression to the mean" has taken place and the slope is somewhat shallower. If there was no correlation, it would be best to simply guess the income mean—as there is no information about income contained in IQ—the red line would be horizontal.

But in neuroscience, things are rarely determined by a single factor.

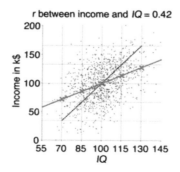

FIGURE 7.3   Regressing income on the basis of IQ, assuming a correlation of 0.42.

If one has multiple predictors, we enter the realm of multiple linear regression, e.g., when we want to predict income from IQ and work ethic (conscientiousness). In that case, we don't fit a line but rather a plane:

$$Y = \beta_0 + \beta_1 X_1 + \beta_2 X_2 + \varepsilon \tag{7.3}$$

Eq. (7.3) is the equation of a plane. We can add as many factors (X_1, X_2, ...) as we want, but we need to estimate a beta-weight for each. Many scientific quests to understand the world we live in (particularly in fields such as economics) boil down to multiple regression problems. If we find a strong beta-weight for work ethic, but not for IQ, we could argue that we live in a world where, in terms of income, hard work matters a lot, but IQ matters little (Fig. 7.4).

In Fig. 7.4, the plane doesn't seem to be such a great fit. Perhaps the influence of IQ on income and work ethic is not independent—perhaps IQ benefits those most who are also willing to work the hardest. This is accounted for by adding an *interaction term* to the equation:

$$Y = \beta_0 + \beta_1 X_1 + \beta_2 X_2 + \beta_3 X_1 X_2 + \varepsilon \tag{7.4}$$

Eq. (7.4) is the same as Eq. (7.3), but with an interaction term. In general, one can add as many factors as one likes, but a "full model" (that includes all variables and interaction effects) has many terms, specifically 2^k terms, where k is the number of factors, so even a 4-factor model already has 16 terms and it is a lot to ask a procedure to estimate all their weights at once.

But we are getting ahead of ourselves. Back to the problem at hand.

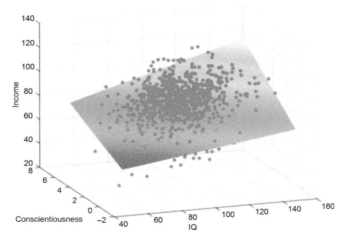

FIGURE 7.4 The relationship between conscientiousness and IQ on income.

The slope and intercept returned by the linear regression tell us the rate at which values of one variable change as those of the other do (the slope), and if we have a y-offset value (the intercept). Here, our x-values are the evokedPSP, and the y-values are the evokedPSTH. In other words, we test how well the instantaneous subthreshold voltage (PSP) predicts the output (PSTH). There are multiple ways of doing linear regression in MATLAB and we'll introduce several of them here. The simplest would be b = evokedPSP' \ evokedPSTH', where we estimate the (correct) beta weight (or slope) by doing a left matrix divide of the two time series as column vectors. This works because one can conceptualize the terms in the regression equation as a system of linear equation: $Y = XB$, where a column vector of y-values is generated by the matrix multiplication of a column vector of x-values and one for weights. A matrix left divide solves for B, where $B = X \backslash Y$. But this is just the slope. We would like to have a model with an intercept. So another way of doing it is by using the function polyfit:

```
p = polyfit(evokedPSP,evokedPSTH,1) %Fitting a polynomial of degree 1
```
p(1) is the slope and p(2) the intercept
```
yPred = polyval(p,evokedPSP); %Evaluating the polynomial on the basis of the evoked PSPs
residuals = evokedPSTH - yPred; %Calculating the residuals as the difference between prediction and actual value
SS_r = sum(residuals.^2); %Calculating the sum of squared residuals
SS_t = var(evokedPSTH).*length(evokedPSTH)-1; %Calculating the total sum of squares
rSquared = 1 - (SS_r/SS_t); %Calculating the degree of variance accounted for as rSquared
```

This will always work. The way that is most analogous to the way we are doing it with the Python stats library is by using the MATLAB function regress from the Statistics and Machine Learning toolbox. Note that, in MATLAB, one indicates placeholders in a function output that one doesn't want to calculate by ~. Here, the inner output parameters of regress are not of interest right now, so we don't calculate them (Fig. 7.5).

Python	MATLAB
| ```
import scipy.stats as st #1
slope1,intercept1,rvalue1,pvalue1,stderr1=st.linregress
(evokedPSP,evokedPSTH) #2
PSPrange = np.linspace(min(evokedPSP),max(evokedPSP),100)
#3
plt.plot(PSPrange,slope1*PSPrange+intercept1,
c='r',lw=2,label='linear fit') #4
plt.scatter(evokedPSP,evokedPSTH,c='k',label='spikes vs
PSP')#5
plt.title('PSTH vs PSP, r2 = '+str(rvalue1**2)[:4])#6
plt.legend(loc='upper left') #7
plt.xlabel('mV');plt.ylabel('spikes') #8
plt.savefig('Figure 7.7.5. spikes vs psp fit.
png',dpi=300)#9
``` | ```
[B,~,~,~,STATS] = regress(evokedPSTH',[ones(size…
(evokedPSP')) evokedPSP']) %2 %10
PSPrange = linspace(min(evokedPSP),…
max(evokedPSP),100); %3
h1 = plot(PSPrange,PSPrange.*B(2)+B(1),…
'color','r','linewidth',2); hold on %4
h2 = plot(evokedPSP,evokedPSTH,'o','color','k',…
'markerfacecolor','k','markersize',6) %5
title(['PSTH vs. PSP, r^2 = ',num2str(STATS(1))]) %6
legend([h1 h2],{'Linear fit', 'Spikes vs. PSP'}) %7
xlabel('mV'); ylabel('spikes'); %8
print('-dpng','-r300','Spikes vs PSP fit'); %9
``` |

(Continued)

Pseudocode

```
1. Import package
2. Perform linear regression between PSP and PSTH
3. Create linear space in range of PSP
4. Plot the regression-fit line
5. Scatter plot the PSTH vs the PSP
6. Set the title
7. Set the legend
8. Set the xlabel
9. Save the figure
%10 In MATLAB, we have to add a vector of ones to get the intercept
```

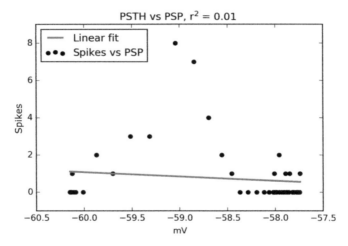

FIGURE 7.5   Predicting PSTHs from PSP. The red line indicates the prediction from the linear regression model. It didn't go so well.

That's a truly terrible fit. In the title of the figure, we plot the r^2 value, which for the relation between PSTH and PSP is equal to 0.01, indicating that the subthreshold voltage accounts for virtually none of the spiking behavior.

This leads us to ask if it is not the *magnitude* of the subthreshold voltage that corresponds to spikes, but rather the *change* in the voltage. That is, we ask not if the instantaneous voltage leads to spikes, but rather the "motion" of the voltage that corresponds to spikes. More specifically, we are interested here in if the *derivative* of the PSP corresponds to spike outputs (Roncek, 1993). While we technically can't calculate the derivative in this discrete space, we can approximate the derivative by calculating the "point-wise" (element by element) difference of the PSP (dPSP), using the function `diff` in POM.

As before, we analyze the relationship between the PSTH and the dPSP by plotting them on the same graph here (Fig. 7.6).

Python	MATLAB
| ```
f, axarr = plt.subplots(1,1,figsize=(6,5)) #1
line1=axarr.plot(evokedRange,evokedPSTH,c='k',label='
PSTH',lw=2)#2
plt.ylabel('spike count')#3
plt.xlabel('time (msec)')#4
ax2 = axarr.twinx()#5
line2 = ax2.plot(evokedRange,evokedDPSP,c='b',
label='dPSP',lw=2)#6
lines = line1+line2#7
lineLabels = [l.get_label() for l in lines]#8
axarr.legend(lines, lineLabels, loc=0,fontsize=10)#9
plt.ylabel('voltage (mV)')#10
plt.title('PSTH and Mean dPSP')#11
plt.savefig('Figure 7.6 - PSTH and mean dPSP.
png',dpi=500)#12
``` | ```
f = figure; %1
[ax, h1, h2] = plotyy(evokedRange,evokedPSTH,…
evokedRange,evokedDPSP); %2 %6 %7 %8
ylabel('spike count') %3
xlabel('time (msec)') %4
title('PSTH and differential of mean dPSP') %11
h1.Color = 'k'; h1.LineWidth = 2;
h2.Color = 'b'; h2.LineWidth = 2;
legend([h1, h2], 'PSTH', 'dPSP'); %9
shg
print('-dpng','-r300',…
'PSTH and differential of mean dPSP'); %12
``` |

Pseudocode

1. Create figure
2. Plot PSTH
3. Set ylabel
4. Set xlabel
5. Double the y-axis to have 2nd axis on right
6. Plot evoked dPSP on 2nd y-axis
7. Hold both lines
8. Get the line labels
9. Set the legend according to line labels
10. Set the ylabel
11. Set the title
12. Save figure

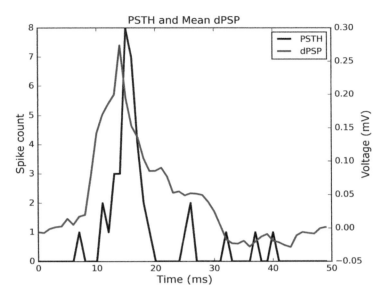

FIGURE 7.6 PSTH (in spikes) vs. change in mean voltage of PSP.

As before, we now perform a linear regression between the PSTH and the dPSP. Recall that an r^2 value of 0 indicates that we cannot account for the changes in the dependent variable by changes in the independent variable with our linear model whereas a r^2 value of 1 indicates that we can perfectly do so. (Fig. 7.7).

Python	MATLAB
| ```
slope2,intercept2,rvalue2,pvalue2,stderr2= st.linregre
ss(evokedDPSP,evokedPSTH) #2
dPSPrange = np.linspace(min(evokedDPSP),
max(evokedDPSP),100) #3
plt.plot(dPSPrange, slope2*dPSPrange + intercept2,
c='r', lw=2,label='linear fit') #4
plt.scatter(evokedDPSP,evokedPSTH,c='k',label='spikes
v dPSP') #5
plt.title('PSTH vs PSP, r² = '+str(rvalue2)) #6
plt.xlabel('mV/ms');plt.ylabel('spikes') #8
plt.legend(loc='upper left') #7
plt.savefig('Figure 7.x. spikes vs dpsp fit.
png',dpi=300) #9
``` | ```
[B2,~,~,~,STATS] = regress(evokedPSTH',…
[ones(size(evokedDPSP')) evokedDPSP']) %2 %10
dPSPrange = linspace(min(evokedDPSP),…
max(evokedDPSP),100); %3
h1 = plot(dPSPrange,dPSPrange.*B2(2)+B2(1),'color',…
'r','linewidth',2); hold on %4
h2 = plot(evokedDPSP,evokedPSTH,'o','color','k',…
'markerfacecolor','k','markersize',6) %5
title(['PSTH vs. dPSP, r^2 = ',num2str(STATS(1))]) %6
legend([h1 h2],{'Linear fit', 'Spikes vs. dPSP'}) %7
xlabel('mV'); ylabel('spikes'); %8
print('-dpng','-r300','Spikes vs dPSP fit'); %9
``` |

(Continued)

Pseudocode

---

1. Import package
2. Perform linear regression between dPSP and PSTH
3. Create linear space in range of dPSP
4. Plot the regression-fit line
5. Scatter plot the PSTH vs the dPSP
6. Set the title
7. Set the legend
8. Set the xlabel
9. Save the figure

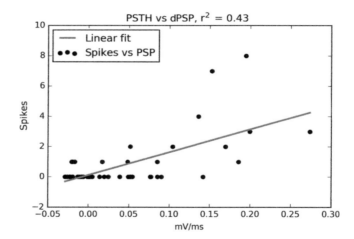

FIGURE 7.7   PSTH predicted from the change in PSP. Somewhat better.

The resulting r^2 value of 0.43 indicates that 43% of the variance in the spiking can be accounted for by the dPSP with our linear regression model. Thus, this linear regression analysis suggests that the derivative of the synaptic input is substantially more correlated with the spiking than the magnitude of the synaptic input. This might be as well as we are going to do here (large ranges of voltage fluctuations seem to fail to evoke *any* spikes), in other words, the relationship is perhaps not all that linear and might include nonlinearities. This is not untypical for neuroscience in general. Overall, the relationship between membrane potentials and spiking behavior is complex, particularly when considering multiple spike trains (Dorn & Ringach, 2003). If you want to know more about how this plays out in auditory cortex, read Nylen (2016). So let's move on to a form of nonlinear regression—logistic regression.

Why Logistic Regression?

Despite all these fancy outcomes that linear regression can bring about, there are limits. Sometimes, as above, there are inherent nonlinearities. Sometimes, binary (or categorical) outcomes are all that you care about. This is surprisingly common in practical applications if you desire to predict win or loss, up or down votes, buy or sell decisions, divorce or mortgage default, depression or cancer, life or death, approach or avoid, stay or go, fight or flight, getting a grant or a job.

What Is Logistic Regression?

It is a kind of simple nonlinear model to link predictors and outcomes, specifically cases where outcomes are categorical—and usually even binary. They are linked through a *sigmoidal* function. It gives the *odds* that an outcome happens— versus it not happening for a given predictor value. As with linear regression, an estimation of beta weights is involved and the fitting of a function (the sigmoidal function) to data. However, getting there is a bit more complex - we have to introduce concepts like odds and the logit function.

What Are Odds?

$$Odds = \frac{p}{\sim p} = \frac{p}{1-p} \tag{7.5}$$

Eq. (7.5): p = probability of an event happening. $\sim p$ = probability of an event not happening. As you can see, odds is simply the technical term for the ratio between the likelihood of an event occurring vs. it not occurring.

How About a Specific Use Case?

Say you want to get into graduate school and want to predict the odds that you will get admitted based on your GRE score (or if you think the great filter is already behind you (Webb, 2002), think of these scores as the number of citations you have, predicting whether you will get a faculty position). If that filter is also already behind you, think of it as predicting whether you will get tenure. If that is also already behind you - well, what do you have to worry about?

TABLE 7.1 Data from previous applicants. Columns: Applicants. Top row: Applicant number, middle row: GRE score, bottom row: Admissions outcome

Applicant	1	2	3	4	5	...	496	497	498	499	500
GRE score	304	279	338	296	299	...	312	290	319	300	293
Admitted	0	0	1	0	0	...	1	0	0	0	0

You have data on previous applicants to the program of your choice specifically you have data from 500 applicants, their GRE scores, and admission outcomes (Table 7.1).

You do what you have learned above—trying to implement linear regression (see Fig. 7.8).

Now what? How is this going to help us? Sometimes, even the best fit line is not a very good fit. This is one of those cases. The relationship is clearly not linear—mostly owed to the fact that the inputs to the model are continuous, but the outcomes are not. There are several problems trying to apply linear regression to this situation. First of all, the assumptions of linear regression such as normally distributed errors are almost certainly violated—this distribution can't be normal if the outcome is dichotomous. Moreover, linear regression is theoretically unrealistic in cases where a threshold is implied. Linear regression assumes that the effect of an unit change in the independent variable on the dependent variable is constant across all values. However, in the admissions case (as in many other cases), changes in the middle of the range affect the probability more dramatically than in the extremes. Most importantly, we are ultimately interested in the probability of getting in with a given score. Probabilities are bounded by 0 and 1. However, linear regression is unbounded. Figure 7.8—mercifully—understates the problem, as we capped drawing it at 0 and 1. But theoretically, the line just keeps on going, which is of course absurd, as probabilities can't be lower than 0 and larger than 1. There are several ways to address these issues, the most commonly used is probably the logit function or logit transform.

What Is the Logit Function?

The logit solves this boundary problem. Instead of looking at the problem in terms of probabilities, the logit looks at log odds. This has two advantages: Whereas probabilities are bounded by 1, odds have no ceiling—they go towards infinity for very high odds. Similarly, whereas probabilities have a floor of 0, taking log odds does not. Taking the log of very small odds (close to 0) will yield values close to negative infinity. So we now have an unbounded variable. In addition, probability does not affect the odds in the same way across the range, so a nonlinearity is built in. Finally, this

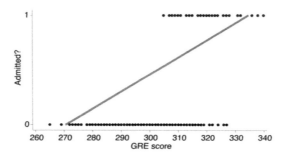

FIGURE 7.8 *x*-Axis: GRE scores. *y*-axis: Admission outcome. Black dots: Individual applicants. Red line: Best fit line as per linear regression.

function is symmetric around the midpoint of $p = 0.5$, so it has a lot of desirable properties for solving this problem. We can then look for beta-coefficients in terms of the logit (on the right hand of equation 7.6). This allows us to look at unit changes of X and how it impacts the logit.

$$\text{logit}(p) = \ln(odds) = \ln\left(\frac{p}{1-p}\right) = \beta_0 + \beta_1 x_1$$

(7.6)

All of This Sounds a Bit Abstract—What Does the Logit Function Look Like?

Like this—see Fig. 7.9.

Are We Done Yet?

Almost. Logit gives us the opposite of what we want. Note that it has probabilities on the *x*-axis. We want them on the *y*-axis. To get there, we have to invert the logit function, which will rotate the graph by 90 degrees. We are interested in probabilities, so we need to solve for p.

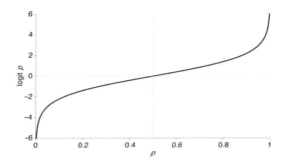

FIGURE 7.9   Not bad. It approaches probabilities of 0 and 1, but never quite reaches them and goes through 0 at $p = 0.5$.

What Does That Look Like?

$$\text{logit}^{-1}(x) = \frac{1}{1 + e^{-x}} = \frac{e^x}{1 + e^x} \tag{7.7}$$

How Does This Help?

We can go back to the definition of the logic, substitute and solve for p:

$$\text{logit}(p) = \ln(odds) = \ln\left(\frac{p}{1-p}\right) = \beta_0 + \beta_1 x_1$$

$$\frac{p}{1-p} = e^{\beta_0 + \beta_1 x_1}$$

$$p = e^{\beta_0 + \beta_1 x_1}(1 - p)$$

$$p = e^{\beta_0 + \beta_1 x_1} - e^{\beta_0 + \beta_1 x_1} * p$$

$$p + e^{\beta_0 + \beta_1 x_1} * p = e^{\beta_0 + \beta_1 x_1}$$

$$p(1 + e^{\beta_0 + \beta_1 x_1}) = e^{\beta_0 + \beta_1 x_1}$$

$$p = \frac{e^{\beta_0 + \beta_1 x_1}}{(1 + e^{\beta_0 + \beta_1 x_1})} \tag{7.8}$$

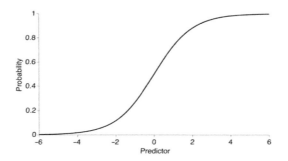

FIGURE 7.10   Probability of outcome given predictor.

What Does That Look Like?

Like this. Note the s-shape of this logistic function. It links predictors (x) to the probability of outcomes (p) (Fig. 7.10).

What Can We Do With That?

Logistic regression estimates the weights that best link predictors to outcomes in a maximum likelihood estimation (MLE) kind of fashion. Logistic regression uses MLE instead of minimizing the squared deviations to estimate the model coefficients. MLE is based on finding the coefficients that are most likely to have produced the data pattern. We can glean the probability of being admitted from the point on the red curve that corresponds to our score on the x-axis (Fig. 7.11).

Let's try this in POM. As usual, we first want to visualize the situation. We use the dataset that contains the data from this graduate school admissions example. It contains two columns: GRE score and admission outcome (0 or 1) from 500 applicants.

But first we need to load the data from the companion website. As it is contained in a .mat file, we need to convert it first. This code produces Fig. 7.12.

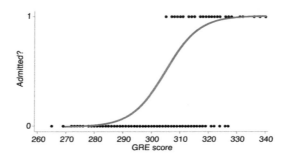

FIGURE 7.11   A logistic fit to the grad school admissions problem, linking GRE scores and admission outcomes. Here, you can read the probability of being admitted with a given GRE score right off the *y*-axis.

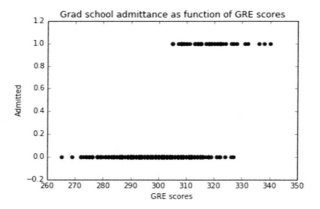

FIGURE 7.12   The dataset of the GRE score as predictors for graduate school admittance. Note low scores generally do not get admitted (0), and high scores do get admitted (1), but there is some overlap, as other factors also play a role. Here, we will use logistic regression to determine the score range for which one is more likely than not to be admitted.

Pseudocode

1. import package to handle mat file
2. initialize list greData
3. open mat file as f
4. for each key in f (there should be only one)
5. for each value in f (there will be two columns)
6. append the value to greData
7. assign x_orig to the GRE scores
8. assign y to the acmittance values
9. create x, GRE values that have been mean-adjusted to zero (since in the standard logistic function in Figure 7.10, y=0.5 when x=0)

Python	MATLAB

```
import h5py #1
greData=[] #2
with h5py.File('logSataSet.mat') as f: #3
    for key in f.keys(): #4
        for vals in f[key]: #5
            greData.append(vals) #6
x_orig = greData[0] #7
y = greData[1] #8
x=x_orig-np.mean(x_orig) #9
```

```
load('logSataSet.mat')
greScore = greAndOutcome(:,1);
admission = greAndOutcome(:,2);
```

Pseudocode

1. plot the data -- admittance vs GRE score
2. set the title
3. label the x-axis
4. label the y-axis

Python	MATLAB

```
plt.scatter(x+np.mean(x_orig),y,c='k') #1
plt.title('Grad school admittance as
function of GRE scores') #2
plt.xlabel('GRE scores') #3
plt.ylabel('Admitted') #4
```

```
plot(greScore,admission,'o','color','k','markerfacecolor','k') %1
title('Grad school admittance as function of GRE scores') %2
xlabel('GRE scores') %3
ylabel('Admitted') %4
```

Pseudocode

---

1. import package for logistic regression
2. create the logistic regression model
3. fit the logistic regression with x and y
4. create a new set of x-values to test, called x_pred
5. predict the admittance for the new x_values
%6 In Matlab, categories have to be positive integers, so we add 1
%7 Determine probabilities of each category with a given score

---

Python	MATLAB

```
from sklearn.linear_model import LogisticRegression #1
logRegression = LogisticRegression() #2
logRegression.fit(x.reshape(len(x),1),y.reshape(len(y),1)) #3
x_pred = np.arange(x.min(),x.max(),1) #4
y_pred=logRegression.predict_proba(x_pred.reshape(len(x_pred),1)) #5
y_proba=np.array([_[1] for_ in y_pred])
```

```
b = mnrfit(greScore,admission+1); %2 %6
probs = mnrval(b,greScore); %7
predictedOutcomes = probs(:,2)>0.5; %5
```

We can visualize this (Fig. 7.13).

Pseudocode

---

1. Plot the new x values vs the predicted admittances, y_pred
2. set the title
3. label the x-axis
4. label the y-axis
5. Plot probability of being admitted with a given score

Python	MATLAB

```
plt.scatter(x_pred+np.mean(x_
orig),y_pred[:,1],c='k') #1
plt.title('Predicted Grad school
admittance as function of GRE
scores') #2
plt.xlabel('GRE scores') #3
plt.ylabel('Admitted') #4
```

```
plot(greScore,probs(:,2),'.','color','b') %5
hold on
plot(greScore,predictedOutcomes,'o','color','k','markerfacecolor','k') %1
title('Predicted grad school admittance as function of GRE scores') %2
xlabel('GRE scores') %3
ylabel('Admitted') %4
```

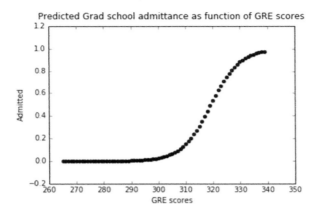

FIGURE 7.13 Logistic regression prediction of admittance. Note that the prediction is that for a GRE score of 320 or higher, it is more likely than not that you will get admitted to grad school, but with a score less than that you will not. Also note that one point difference in GRE score makes the most difference in terms of admissions probability around that value. In other words, the impact of a single point change on probability is not constant. This makes linear regression less and logistic regression more suitable for this problem.

This Is Still Too Abstract. Can We Apply This to Something More Neural?

Let's try it. How about perceptual neuroscience? Say we want to predict whether someone saw "the dress" as white and gold or differently and we have a large dataset with some predictors? This problem basically screams "logistic regression." Let's build a model.

In February 2015, an image of a wedding dress made a big splash on social media. The reason for this was that this image was the first one in the color domain that was not only ambiguous, but also lent itself to profoundly differing interpretations, e.g., seeing it as white and gold, black and blue, or other colors altogether (Fig. 7.14). The reason we are discussing it here is that it is also interesting for our understanding of color vision (Wallisch, 2015). The light impinging on our visual system is a mixture of the wavelength from the illumination source as well as the object. In a process called "color constancy," the visual system has to discount the wavelengths from the illumination source, so that the same object looks the same, regardless of illumination (Wallisch, 2016). This is a tricky problem if the illumination source is not well defined, as in the case of the dress. In that case, people might go with their illumination priors. For instance, if someone is a "night owl" and exposed to more long-wavelength lights from incandescent light sources, they might assume that the illumination was yellowish-artificial light. Mentally subtracting that from an ambiguous image might

FIGURE 7.14   Kira Wallisch wearing "the dress."

yield a black and blue percept. In contrast, if someone is a "lark" and rises with the sun, they are exposed to relatively more short wavelengths from the sun. If that is so, they might assume natural lighting and mentally subtract blueish light from the ambiguous image, yielding a white and gold percept (Wallisch, 2016). We collected logged data to see if this is the case. For this exercise we share data from 5000 participants here. Each row represents the responses of a participant. The first column indicates what they saw, the second column whether they thought the dress was in a shadow when the picture was taken, the third column represents age in years, and the fourth column is self-identified circadian type. As the data are mixed numbers and characters, we use a cell array to represent these data.

So let's write the dress code. We are building it up in the fashion of our classical canonical data processing cascade that we introduced in Chapter 4, Correlating Spike Trains.

First, we initialize and write the loader, steps 0 and 1 of the cascade.

Pseudocode

1. Clear the workspace
2. Close all the figures
3. Clear the screen
4. Load the workspace with the dress data
5. Determine the number of participants
6. Import Python package h5py (great for loading .mat files)
7. Initialize dressData
8. Load file with h5py.File, use as the variable f
9. For each row in f, initialize row_data
10. For each row in column
11. Try to append
12. Convert the numbers contained in f[column[rowNumber]] to characters with chr(), join each character with the function join()
13. Except if there is an error in the chr() string conversion (that is, if it tries to convert a NaN)
14. Just add an empty string
15. Append the rowData to dressData
16. Transpose the Python dressData to make parallel to MATLAB code

Python	MATLAB

```
import h5py # 6
dressData = [] # init
with h5py.File("dressData.mat") as f: #8
    for column in f['DATA']: #9
        rowData = [] #9
        for rowNumber in range(len(column)): #10
            try: #11
                rowData.append(''.join([chr(_) for _ in
f[column[rowNumber]][:]])) #12
            except: #13
                rowData.append('') #14
        dressData.append(rowData) #15
dressData=np.array(dressData).T #16
numParticipants=len(dressData) #5
```

```
clear all %1
close all %2
clc %3
load('dressData.mat') %4
numParticipants = length(DATA);
%5
```

Before we can do anything else, we need to eliminate data from people who did not respond to the question how they saw the image of the dress. These data will be useless—the earlier we eliminate them from the analysis, the better. So on to step 2, the pruner.

Pseudocode

---

1. Initialize a variable that will contain information about whether a participant responded to the perceptual question
2. Determine whether the perceptual response of a given participant was a nan
3. Do this for all participants
4. Find indices of those that didn't respond
5. Eliminate them from the dataset
6. Update the number of participants—count only the valid ones

---

Python	MATLAB
| ```
nonResponder = np.zeros(numParticipants) #1
killSet=[];
for nn in range(numParticipants): #3
    if dressData[nn][0]=='': #2
        killSet.append(nn) #4
dressData = [v for i, v in enumerate(dressData) if i
not in killSet] #5
numParticipants = len(dressData) #6
``` | ```
nonResponder = zeros(numParticipants,1); %1
for nn = 1:numParticipants %3
    nonResponder(nn,1) = sum(isnan(DATA{nn,1})); %2
end %3
killSet = find(nonResponder==1); %4
DATA(killSet,:) = []; %5
numParticipants = length(DATA); %6
``` |

We are left with 97% of the data—for data that were logged online (the net is teeming with trolls), this is not a bad response rate.

On to step 3—the parsing of the data. We'll use this as an example of how to tackle string handling in POM. The data are in the form of strings—people responded to what they saw. In order to do computations on that, we'll have to transform that into numbers.

Pseudocode

1. Preallocate a matrix of suitable length that will contain the parsed numerical responses with zeros
2. Parse the perceptual response of a given participant and assign a numerical code, from 1 to 5
3. Parse the circadian type of a given participant and assign a numerical code, from 1 to 5
4. Loop through all participants

Python	MATLAB

```python
numResponses = np.zeros(shape=(numParticipants,2)) #1
for nn in range(numParticipants): #4
    if dressData[nn][0]=='White/Gold': #2
        numResponses[nn][0] = 1 #2
    elif dressData[nn][0]=='Blue/Black': #2
        numResponses[nn][0] = 2 #2
    elif dressData[nn][0]=='Blue/Gold': #2
        numResponses[nn][0] = 3 #2
    elif dressData[nn][0]=='White/Black': #2
        numResponses[nn][0] = 4 #2
    else:
        numResponses[nn][0] = 5 #2
    if dressData[nn][3]=='Strong owl': #3
        numResponses[nn][1] = 1 #3
    elif dressData[nn][3]=='Owl':#3
        numResponses[nn][1] = 2 #3
    elif dressData[nn][3]=='Lark':#3
        numResponses[nn][1] = 3 #3
    elif dressData[nn][3]=='Strong lark': #3
        numResponses[nn][1] = 4 #3
    else: #3
        numResponses[nn][1] = 5 #3
```

```matlab
numResponses = zeros(numParticipants,2); %1
for nn=1:numParticipants %4
    if strcmp(DATA{nn,1},'White/Gold') == 1 %2
        numResponses(nn,1) = 1; %2
    elseif strcmp(DATA{nn,1},'Blue/Black') == 1 %2
        numResponses(nn,1) = 2; %2
    elseif strcmp(DATA{nn,1},'Blue/Gold') == 1 %2
        numResponses(nn,1) = 3; %2
    elseif strcmp(DATA{nn,1},'White/Black') == 1 %2
        numResponses(nn,1) = 4; %2
    else %2
        numResponses(nn,1) = 5; %2
    end %2

    if strcmp(DATA{nn,4},'Strong owl') == 1 %3
        numResponses(nn,2) = 1; %3
    elseif strcmp(DATA{nn,4},'Owl') == 1 %3
        numResponses(nn,2) = 2; %3
    elseif strcmp(DATA{nn,4},'Lark') == 1 %3
        numResponses(nn,2) = 3; %3
    elseif strcmp(DATA{nn,4},'Strong lark') == 1 %3
        numResponses(nn,2) = 4; %3
    else %3
        numResponses(nn,2) = 5; %3
    end %3
end %4
```

We are now in a position to proceed to step 4a, the exploratory data analysis. We always recommend to do this step first, in order to see whether the data, at first glance, make sense. If you know what to expect (e.g., from the literature), this will also give you an idea of whether the parsing step worked as intended.

III. GOING BEYOND THE DATA

Pseudocode

---

 1. Define an edges vector that we'll use to parse the data
 2. Compute marginals for both kinds of numerical responses, converted to percentages
 3. Allocate suitable tick labels
 4. Open a new figure
 5. Open a new subplot
 6. Make a bar plot of the marginals over the edges vector
 7. Change color of bars to a light shade of grey
 8. Fix figure aesthetics, turn tick direction outwards, remove box, change font
 9. Add axis labels and format them properly
10. Do this for both subplots
11. Make the layout tight

Python	MATLAB

```python
edges = range(int(numResponses.min()),int(numResponses.
max())+1) #1
edgesplus = range(int(numResponses.min()),int(numResponses.
max())+2) #1
marginals = []
marginals.append(np.histogram(numResponses[:,0],edgesplus)
[0]) #2
marginals.append(np.histogram(numResponses[:,1],edgesplus)
[0]) #2

TickLabels = [['W/G','B/B','B/G','W/B','O'],\
['SO','O','L','SL','O']]#3

f = plt.figure; #4
fsize=16; #8
for ii in range(2): #10
    ax = plt.subplot(1,2,ii+1); #5
    h = plt.bar(edges,marginals[ii]/float(numParticipants)*10
0,facecolor=[.9,.9,.9]); #6
    plt.ylabel('Percentage',fontsize=fsize,style='italic') #9
    plt.xlabel('Type',fontsize=fsize,style='italic') #9
    plt.xticks([edge+.5 for edge in edges],TickLabels[ii])
plt.tight_layout()
```

```matlab
edges = [min(unique(numResponses)):…
max(unique(numResponses))]; %1
marginals(1,:) = (histc(numResponses(:,1),…
edges)./numParticipants).*100; %2
marginals(2,:) = (histc(numResponses(:,2),…
edges)./numParticipants).*100; %2

TickLabels(1,:) = {'W/G','B/B','B/G','W/B','O'}; %3
TickLabels(2,:) = {'SO','O','L','SL','N'}; %3
f = figure; %4
for ii = 1:2 %10
ax = subplot(1,2,ii); %5
h = bar(edges,marginals(ii,:)); %6
set(h,'facecolor',[0.9 0.9 0.9]); %7
ax.FontSize = 16; ax.FontAngle = 'italic'; %8
ax.TickDir = 'out'; box off; %8
ylabh = get(gca,'YLabel'); %9
ylabh.Position(1) = 0; %9
ax.XTickLabel = TickLabels(ii,:); %9

ylabel('Percentage') %9
xlabel('Type') %9
end %10
```

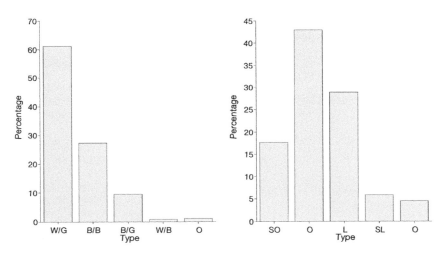

FIGURE 7.15 Marginal distributions of numerical responses.

This should yield Fig. 7.15, showing the marginal distributions of the numerical responses. They make sense. Most people report a W/G percept, consistent with early reports on social media and more people report to be owls than larks, with extreme types less likely than moderate ones.

Just to be safe, we also do a preliminary analysis to see if there is even anything to explain—like a basic relationship between circadian type and response, before moving on to a full blown regression model. So on to step 4b.

Pseudocode

```
1. Find indices of numerical responses that corresponding to a given circadian type
2. Use those to find the perceptual responses of all participants with that circadian type
3. Find the proportion of those that reported white and gold and convert to percentages
4. Do this for all valid circadian types (1 to 4)
5. Open a new figure
6. Make a bar graph of the percentages as a function of circadian type
7. Set proper ylimits to discern whether there is an effect
8. Add and format axis labels
9. Add string tick labels on the x-axis
10. Make formatting nice—increase fontsize, change direction of tick marks, remove the box
```

(Continued)

Python	MATLAB
| ```python
RESULT=np.zeros(4)
for ii in range(4): #4
    temp = np.where(numResponses[:,1]==ii+1) #1
    ColorSubSet = numResponses[temp,0] #2
    RESULT[ii] = len(np.where(ColorSubSet[0]==1)[0])/float(len
(ColorSubSet[0]))*100
f = plt.figure() #5
plt.bar(range(4),RESULT) #6
plt.ylim([59, 66]) #7
plt.xlabel('Circadian type') #8
plt.ylabel('Percent seeing white and gold') #8
plt.xticks([edge-.5 for edge in edges[:-1]],TickLabels[1]
[:-1])
``` | ```matlab
for ii = 1:4 %4
    temp = find(numResponses(:,2)==ii); %1
    ColorSubSet = numResponses(temp,1); %2
    RESULT(ii,1) =...
    (length(find(ColorSubSet==1))./...
    length(ColorSubSet)).*100; %3
end %4

figure %5
bar(1:4,RESULT) %6
ylim([59 66]) %7
xlabel('Circadian type') %8
ylabel('Percent seeing white and gold') %8
ylabh = get(gca,'YLabel'); %8
ylabh.Position(1) = 0.3; %8
set(gca,'XTickLabel',TickLabels(2,:)); %9
set(gca,'Fontsize',20) %10
box off; set(gca,'tickdir','out') %10
``` |

This yields Fig. 7.16.

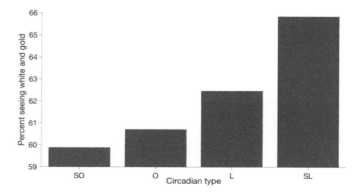

FIGURE 7.16   Percent seeing white and gold as function of circadian type. There seems to be a relationship between circadian type and the probability of seeing the dress in a certain way, and in the direction predicted by our color constancy theory! Reassuring.

Looks like there is a relationship between self-identified circadian type and self-reported percept, in the direction we expected on the basis of theoretical considerations. The effect is modest, but seems robust—not bad, given that these data were logged online, where lots of noise is to be expected.

Now, we are in a position to perform a logistic regression on the dress data.

As we have only coarse information from four discrete categories and even the lowest one (strong owls) has a probability of almost 60% of seeing white and gold, we have to take other information into account. Luckily, we have this information available, such as whether participants believe the object to be in a shadow or not. Shadows have blue spectral wavelength content. Subtracting a shadow prior from an ambiguous image should boost the probability of seeing the stimulus as white and gold. On the companion website, you can download a program that implements a model with multiple factors, and then does multiple logistic regression. There are other factors as well, such as age—it might matter how long one has been exposed to natural or artificial light, determining the strength of these priors.

Beyond 1: Being able to do and understand linear and logistic regression is a pretty good point to be in terms of going from 0 to 1. Of course, there are many other kinds of regression beyond 1. As data based prediction is a key component of machine learning and data science applications, this is a thriving area of research. One method that we would like to name here explicitly is "regularization" (Figs. 7.17 and 7.18).

Regularization

Regularization is a bit beyond the scope of the book (well beyond 1, an area of active research), but it is worthwhile to give an intuition of what people talk about when talking about regularization, at you will likely encounter it. In the regression procedures above, we end up with an equation that is useful for predicting some output, given some inputs. As we've seen, all these equations have coefficients (weights/parameters) and the methods are designed to give you the "best fit," the optimal way to relate the predictor variable(s) to the outcome variable. However, as we have seen repeatedly, this can be problematic, as these methods calculate a best fit given a particular dataset, which is usually called the "training dataset" (more on this in chapter: Classification and Clustering). Overfitting is a constant threat. In other words, the "best fit" for a given dataset is ironically not likely to be the best fit in general. This is a concern as it can leave us overconfident in our understanding of the world. It is a particular concern when the number of coefficients to be estimated from a given dataset is large, to fit complex models. Regression methods will do it; but how reliable are the coefficients that we estimate? Having complex models is, in principle, a good thing, as there will always be a lot going on in neuroscience data and we want the model to do justice to that complexity. But there is always a lot of (measurement) noise present, so we need to make sure that the model doesn't just fit the noise—because that won't (by definition) generalize. That is where regularization methods come in, they push back against overfitting a bit, balancing the least squares solution. The basic idea is that models with less extreme beta weights are more likely to generalize, so models that have

FIGURE 7.17   DATA.

FIGURE 7.18   SATA.

coefficients that are too large are discouraged. In practice, a "regularization term" that penalizes large terms ("shrink-age") is added to the equation to give simpler models a fair chance of being the "best" model. Often, this leads to a "smoothing" of the regression line, as sharp turns are abhorred by nature and are more likely to be indicative of overfit-ting to noise. Of course, this introduces something of a fudge factor—how much (and how, specifically) should complex-ity be penalized? As a consequence, there are several different specific regularization methods, including the popular *LASSO* and *Ridge* regressions (also known as L1 and L2 regulatization), which can have different impacts on your pre-dictions. Lasso (or "L1 norm") methods take the sum of the absolute value of the coefficients as a regularization term whereas Ridge (or "L2 norm") methods take the sum of the squared coefficients. To summarize, in multiple regression, it is always hard to tell which coefficients to use (if their number is large) and which ones to turn off. Instead of doing that, including a regularization term in the regression allows for a smoother selection criterion - perhaps making all coefficients a bit smaller, which tends to yield more stable solutions.⌘

⌘. *Pensee on the limits of prediction by extrapolation:* In the early 1970s, the Club of Rome commissioned a report on the future of humanity based on computer simulations of population growth, economic growth, and the availability of natural resources, titled "The limits to growth" (Meadows, Meadows, Randers, & Behrens, 1972). It was an immediate bombshell upon publication, with millions of copies sold and having a deep impact on public discourse—a lot of their predictions becoming "conventional wisdom." They predicted that within a few decades, humanity would run out of all essential natural resources such as gold, silver, copper, oil, aluminum, mercury, and zinc, bringing about societal collapse in the relatively near future. A more appropriate title would have been "The limits of predicting the future on the basis of extrapolating computer models." None of the dire predictions in the book were even close to coming to pass (Simon, 1998; Lomborg, 2003). Known reserves of all seven commodities mentioned above are now more abundant than they were in 1972 (and for the most part cheaper as well) well past the time when we were supposed to have run out of all of them (in their most optimistic scenario, they assumed 5× known reserves, projecting the last one (aluminum) to run out by 2027; in reality, bauxite [the principal source of aluminum] production and known reserves are at all time highs). Similarly, food production had no problem keeping up with increased population and neither have we run out of farmland. Was this just an alarmist doomsday scenario peddled by the Club of Rome in an attempt to get attention or were they sincerely concerned? Even though it was probably intended to be a warning to humanity to fundamentally change its ways or suffer the consequences, it really serves as an (underappreciated) warning about the inherent perils of prediction by extrapolation. All real-life social and biological systems have so many feedback loops—to say nothing of the role of innovation or the impact of black swan events (Taleb, 2007)—that any ballistic prediction is likely to be misleading in the extreme. If you know of someone who does this in business, please let us know so we can bet against them. Simply extrapolating a complex and poorly understood system is a bad idea, always. At this point, Malthus has been wrong for almost 200 years (Malthus, 1826). Obviously this is not stopping neo-Malthusians from attempting similar folly. The fact that we don't learn from apocalyptic predictions that don't come to pass suggests that there is tribalist circuitry at work that prevents us from doing so. In other words, some of us *want* the world to end and are looking to science for their belief that they live in the endtimes, all the evidence to the contrary notwithstanding. Pessimism is oddly popular and even seen as a sign of sophistication. But it is not edgy to be wrong. It is just wrong. Also, lest we be accused of being partial, this is about intellectual integrity, not picking sides in ideological battles. The other side—expecting an imminent technological rapture (the "singularity" (Kurzweil, 2005))—for no other reason than that it would be highly desirable is likely to be equally disappointed by the future. Historically, the predictions made by pollyannish cheerleaders on the basis of an extrapolation of positive exponential trends also have an extremely poor track record of coming true, although they tend to have the positive effects of innovation on their side (Kurzweil, Richter, & Schneider, 1990). It is extrapolation itself that is the problem; doing so exponentially just magnifies the likely error. If you do this, the best you can hope for is for people to have short memories. Regression works best if it relates the states of two stable systems, but complex systems are rarely stable over time, which makes extrapolating over time particularly challenging.

III. GOING BEYOND THE DATA

Dimensionality Reduction

Some sciences like physics or chemistry are simple because there are many causal relationships involving a few linearly decomposable factors that almost completely determine outcomes. A prominent example of this is the functional dependence of reaction rates on temperature—the rate of a chemical reaction is an exponential function of temperature—plotted in log space, the relationship is linear—it can be expressed by a single line, and that is pretty much all there is to it (Arrhenius, 1889a, 1889b).

Neuroscience is not like that. There is no fundamental unit of analysis that is simple and elementary enough to lend itself to analysis in isolation. The concept of a "simple system" is a misnomer in neuroscience, as even the simplest systems are inherently complex. Many numbers are needed to reasonably characterize a system, such as a network of neurons or even a single neuron. In other words, methods to deal with multivariate data are essential. One key method to tackle something as involved as a network of neurons is dimensionality reduction.

To wit, have a look at this multivariate dataset that involves only 14 repeated measures of 12 variables (Fig. 8.1). Can you discern the underlying structure?

As primates, we can maximally visualize things in three or four dimensions, but as we record from more and more neurons at once in order to characterize a neural network more fully, methods to transform a many-dimensional dataset into one that can be represented by three or four dimensions without losing too much information are becoming ever more important.

This is a good point to introduce a second code design principle. In Chapter 4, Correlating Spike Trains, we detailed how analysis code could (and perhaps should) be organized, namely in an analogy to perceptual systems. We called this the "canonical processing cascade." Here, we want to introduce the concept of "levels of coding," in analogy to the primate motor system (Fig. 8.2).

Neural Data Science.
DOI: http://dx.doi.org/10.1016/B978-0-12-804043-0.00008-8

	1	2	3	4	5	6	7	8	9	10	11	12
1	8	6	8	7	8	2	1	8	4	1	3	5
2	5	1	7	6	1	6	4	4	1	2	8	4
3	9	1	8	9	7	6	2	4	3	5	8	9
4	3	8	2	3	9	5	8	4	7	8	1	3
5	7	1	8	5	3	1	9	2	4	6	2	4
6	6	9	5	2	9	2	4	5	5	2	3	4
7	3	2	1	8	8	3	4	5	1	7	8	7
8	1	4	5	7	9	6	3	5	6	7	9	5
9	8	3	4	8	5	3	9	4	7	4	2	1
10	1	5	7	8	5	6	1	2	2	9	2	6
11	4	4	6	7	8	8	4	9	8	1	1	4
12	8	8	4	6	1	3	1	8	6	3	5	2
13	8	1	2	4	5	5	1	2	3	6	6	3
14	1	8	9	6	2	3	1	2	6	8	6	2

FIGURE 8.1   Multivariate dataset.

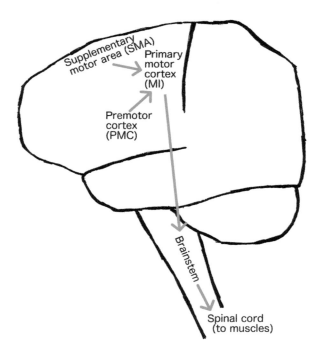

FIGURE 8.2   The primate motor system. Note that this is a cartoon. There are many other structures involved, that modulate and gate motor behavior, e.g., the cerebellum and the basal ganglia, but also others, such as prefrontal cortex. The core areas are shown here though.

The motor system is in the business of telling the muscles what to do. As a coder, you are in the business of telling a computer what to do. It is advisable to glean some of the basic design principles from the motor system and port it to coding principles.

The motor system is organized in terms of three hierarchical levels, with increasing degrees of freedom as one goes down the hierarchy. Conceptually, the top level consists of formulating a strategic goal—something fairly high level that you want to achieve, say you want to throw a ball into a net. In the brain, this level is implemented in the premotor cortex (PMC) and the supplementary motor area (SMA). These areas project to neurons in primary motor cortex (MI). The job of this area is to translate the strategic goal into tactics. Importantly, there are many ways in which one could translate the general goal into spatio-temporal activity patterns, for instance one could attempt an overhanded or an underhanded throw. It is the function of training and practice to translate the general movement goal into reliable patterns of spatio-temporal activations in motor cortex that will then translate to spatio-temporal patterns of joint movements and muscle activations (Churchland, Byron, Ryu, Santhanam, & Shenoy, 2006). I might have the same athletic aspirations as a professional athlete, but without the practice to translate them into reliable tactics, I have no hope of achieving that. In turn, the primary motor cortex talks to the brainstem, which translates these commands into specific muscle activations or outputs via the spinal cord. This corresponds to the execution or implementation part of the hierarchy. Like in the previous part, the number of degrees of freedom increases, as one goes down the hierarchy; there are many combinations of muscle activations that can, in principle, implement the same overall tactic. Perhaps another function of training is for the brain to figure out which such combination is the most efficient one. You can experience something like this by trying to write your name with your nondominant hand (or other extremity). You have the same strategic goal—writing your name—but the execution will be pitiful (though it will improve with practice).

In analogy, code can be efficiently organized in this manner. In neural data science, the strategic goal is the computational task that you want to solve. In this book, each chapter has its own strategic goal. In this chapter, our strategic goal will be dimensionality reduction of a multivariate *sataset*. The next level in coding (the tactical level) is the algorithmic level. There are many algorithms that can implement dimensionality reduction, such as principal component analysis (PCA), independent component analysis (ICA), or linear discriminant analysis (LDA), among many others. Which one is most suitable depends on the structure of the dataset. You could write a function or code section to implement one such algorithm. Finally, the third level is the implementation of the algorithm in code in POM (or another language, if you so choose). The actual code that implements the algorithm consists of instructions that are executed by the computer. Again, you have tremendous degrees of freedom to implement the algorithm, not just in the way you name your variables, but also whether you want to use for loops or vectorized code, which specific subfunctions to use and so on. Just like life, coding is a veritable garden of forking paths—it matters which one you go down (Borges, 1962).

It is important to keep these different levels in mind when organizing your code as well as when discussing and reconciling it with colleagues (who might have made different choices lower in the hierarchy). We suggest to also use this for commenting. The strategic goal of a file should be stated right in the header. The algorithm used should be stated in a comments section that begins a code section or function (noting why that algorithm was used). The implementation should be commented by code pointers, one per line, corresponding to a block of pseudocode nearby. This will make it much more easy to scrutinize your code in light of the many, often arbitrary, choices of implementation at the lower levels of the hierarchy—try looking at your own code after some time with and without this structure. You will see a clear difference in your ability to discern what is going on.

Let's jump into implementing our strategic goal for this chapter- dimensionality reduction- with a specific algorithm, PCA.

Like several other things that we've encountered in previous chapters, the fundamental idea behind PCA amounts to a profoundly platonic enterprise. The basic idea is that the messy and complex world we observe is essentially a mirage and comes about from the (linear, in the case of PCA!) interplay of a few latent (i.e., hidden) factors that live in the world of ideas. However, it is these factors and their interactions that underlie everything and drive things in the real world (plus some noise, to make it less obvious what is going on). The point of PCA is to use data recorded in the real world to run this process in reverse—to uncover the latent factors that brought the data about, which will allow us to understand the world at a deeper level.

If this sounds too abstract, perhaps an example will be instructive. One of the central tenets of contemporary personality psychology is that the personality of an individual can mostly be decomposed into five stable factors: Emotional stability (neuroticism), Extraversion, Openness, Agreeableness, and Conscientiousness (Costa & McCrae, 1992; McCrae & Costa, 1997). This model came about by doing a PCA on the responses of large samples of participants to various personality questionnaires (Costa & McCrae, 1985; McCrae & Costa, 1987). In this framework, the personal idiosyncracies of individuals are manifestations of how the levels of the underlying five factors (plus some noise) combine in any given person (McCrae & Costa, 2003).

How does it work? We will walk you through the process of how to do PCA by doing it on a sataset of neural data in POM.

This sataset represents properties of neurons that could be feasibly measured: response latency, interspike interval (ISI), somatic volume, cortical depth, maximum firing rate, spontaneous (baseline) firing rate spike waveform width, axon length, and dendritic arborization area. In this example, we simulate these values for six different types of neurons, where the mean and standard deviations of each property are used to generate values. Of the six different neural types, three are considered inhibitory and three are excitatory. We will be using dimensionality reduction to reduce the eight dimensions of sata down to a lower number of dimensions.

So let's generate some sata.

Pseudocode	Specify the cell fields we'll be using: latency, isi, volume, z, prob, spont, waveWidth, axon, dendrite
	Specify the fields to generate the sata, namely, means and standard deviation values for the cellField values, ordered in this manner
	Specify each property value (means and standard deviations) for each cell type (1–6)
	Convert to a dataframe

Python

```python
import pandas as pd

cellFields = ['latency','volume','z','maxrate','spont','waveWidth','axon','dendrite']
generatorFields=['Type', 'transmission', 'latMean','latStd','volMean','volStd', 'zMean', 'zStd',
'maxrateMean', 'maxrateStd','spontMean','spontStd','waveWidth','waveStd','axonMean','axonStd',
'dendriteMean','dendriteStd']

Type1=[1, 1, 14, .5,  150,30,500,20,.9, .1,.02,.01, 1, .05, 160, 20, 180,30 ] #L4 pyramid
Type2=[2, 1, 15, .5,  120,30,300,20,.8, .1,.07,.01, 1, .04, 150,  20, 150,30  ] #L2 pyramid
Type3=[3, 0, 15, 1,   120,30, 300,20, .95,.1,.2,.1, .2,.001,150,  10, 150,10  ]  #L2 inhibitory
PV
Type4=[4, 0, 17, 4, 110,30,300,20, .3, .1,.02,  .01,.3,.005,150,  10, 150,40  ]  #L2 inhibitory
Som
Type5=[5, 1, 22, 5,  180,20,800,100,.35,.2,.35, .1,.5,.1,  1000, 500, 200,60 ]  #L6 excitatory
pyramid
Type6=[6, 0, 14.5, .5, 100,30,500,20,.95,.1,.2,.1, .2,.001,150,  10, 150,10  ]  #L4 inhibitory FS

dftype = pd.DataFrame([Type1,Type2,Type3,Type4,Type5,Type6],columns=generatorFields)
```

MATLAB

```matlab
cellT = zeros(6,18);

cellT(1,:)=[1,1,14,.5,150,30,500,20,.9,.1,.02,.01,1,.05,160,20,180,30];
cellT(2,:)=[2,1,15,.5,120,30,300,20,.8,.1,.07,.01,1,.04,150,20,150,30];
cellT(3,:)=[3,0,15,1,120,30,300,20,.95,.1,.2,.1,.2,.001,150,10,150,10];
cellT(4,:)=[4,0,17,4,110,30,300,20,.3,.1,.02,.01,.3,.005,150,10,150,40];
cellT(5,:)=[5,1,22,5,180,20,800,100,.35,.2,.35,.1,.5,.1,1000,500,200,60];
cellT(6,:)=[6,0,14.5,.5, 100,30,500,20,.95,.1,.2,.1,.2,.001,150,10,150,10];

cellT = array2table(cellT,'VariableNames',{'Type', 'transmission',…
'latMean','latStd','volMean','volStd', 'zMean', 'zStd', 'maxrateMean',…
'maxrateStd','spontMean','spontStd','waveWidth','waveStd','axonMean','axonStd',…
'dendriteMean','dendriteStd'});
```

III. GOING BEYOND THE DATA

Note that we use the concept of *DataFrames* here. In Python, this is done with the package *pandas*. DataFrames are two-dimensional data structures that are suitable for tabular data, similar in organization to a spreadsheet, like Microsoft Excel. The organizing principle of the DataFrame is that data stored in it can be assigned a column label, whereas multiple instances of the same unit of analysis (participants, neurons, animals, etc.) can be stored in rows of the DataFrame. This is a great way to store and organize big and complex datasets if you are normally used to storing values in arrays but trust yourself to just remember what the 12th column of data represents—which you probably won't, shortly after writing the code. Generally speaking, DataFrames make it easier for humans to interact with (complex) high-dimensional data.

It would be natural for us to simply represent this sata in MATLAB as a matrix as it is generally advisable to do—the entire environment is optimized to dealing with them, which is why we did that in Chapter 3, Wrangling Spikes Trains. We could simply keep track of what columns and rows represent. Here rows represent neuron types whereas columns represent neural properties. However, since MATLAB 2013b, MATLAB now offers a data type that corresponds to a data frame, namely a "table." For educational purposes, we will here introduce the notion of a table in MATLAB. The reason a table is called that comes from relational databases such as mySQL, which store data in tables. If you know SQL, most of the operations you are familiar with from there, such as selecting all entries that meet certain criteria or performing inner joins carry over to MATLAB tables.

Now that we defined the six cell classes and their respective properties, let's create 100 neurons, randomly picked from these six cell types.

Pseudocode
1. Initialize `sataset`, containing values of each cell type
2. Initialize `celltypes`, which will be 1 and 0 for excitatory and inhibitory
3. For ii in 100 neurons
4. Randomly select a cell type (1–6) to simulate properties and call it `tt`
5. Simulate property `latency`, where `ix` is the pandas indexing function, so that it grabs the mean latency (`latMean`) for cell type `tt` and adds some noise as determined by `latStd` times a randomly drawn number
6. Simulate each other property for the neuron
7. Append each simulated property for this neuron to the `sataset`, where value `ignore_index=True` tells pandas to create a new index for each appended dataframe (otherwise, the index value of each additional cell will be 0. Try using `ignore_index=False` to see this in action)
8. Append the cell type to separate dataframe `celltypes` to use as training data

---

(Continued)

```
Python     import numpy as np
           sataset = pd.DataFrame(columns=cellFields) #1
           celltypes = pd.DataFrame(columns=['transmission']) #2
           for ii in range(100): #3
               tt = np.random.randint(6) #4
               trans = dftype.ix[tt,'transmission'] #5
               latency = dftype.ix[tt,'latMean']+np.random.randn()*dftype.ix[tt]['latStd'] #5
               vol=dftype.ix[tt,'volMean']+np.random.randn()*dftype.ix[tt]['volStd'] #6
               z = dftype.ix[tt,'zMean']+np.random.randn()*dftype.ix[tt]['zStd'] #6
               maxrate = dftype.ix[tt,'maxrateMean']+np.random.randn()*dftype.ix[tt]['maxrateStd'] #6
               spont = dftype.ix[tt,'spontMean']+np.random.randn()*dftype.ix[tt]['spontStd'] #6
               waveWidth = dftype.ix[tt,'waveWidth']+np.random.randn()*dftype.ix[tt]['waveStd'] #6
               axon = dftype.ix[tt,'axonMean']+np.random.randn()*dftype.ix[tt]['axonStd'] #6
               dendrite = dftype.ix[tt,'dendriteMean']+np.random.randn()*dftype.ix[tt]['dendriteStd'] #6

               sataset=sataset.append(pd.DataFrame
               ([[latency,vol,z,maxrate,spont,waveWidth,axon,dendrite]],columns=cellFields),ignore_index=True) #7
               celltypes = celltypes.append(pd.DataFrame([[trans]],columns=['transmission']),ignore_index=True) #8

MATLAB     numNeurons = 100;
           numCellTypes = size(cellT,1);
           sataSet = table(0, 0, 0, 0, 0, 0, 0, 0);
           cellTypes = table(0,0);
           sataSet.Properties.VariableNames = {'latency','vol','z','maxrate','spont', 'waveWidth', 'axon', ...
           'dendrite'};
           cellTypes.Properties.VariableNames = {'transmission','Type'};

           for nn = 1:numNeurons
               cellClass = randi(size(cellT,1));
               cellTypes.transmission(nn,1) = cellT.transmission(cellClass);
               cellTypes.Type(nn,1) = cellClass;
               sataSet.latency(nn,1) = randn(1).*cellT.latStd(cellClass) + cellT.latMean(cellClass);
               sataSet.vol(nn,1) = randn(1).*cellT.volStd(cellClass) + cellT.volMean(cellClass);
               sataSet.z(nn,1) = randn(1).*cellT.zStd(cellClass) + cellT.zMean(cellClass);
               sataSet.maxrate(nn,1) = randn(1).*cellT.maxrateStd(cellClass) + cellT.maxrateMean(cellClass);
               sataSet.spont(nn,1) = randn(1).*cellT.spontStd(cellClass) + cellT.spontMean(cellClass);
               sataSet.waveWidth(nn,1) = randn(1).*cellT.waveStd(cellClass) + cellT.waveWidth(cellClass);
               sataSet.axon(nn,1) = randn(1).*cellT.axonStd(cellClass) + cellT.axonMean(cellClass);
               sataSet.dendrite(nn,1) = randn(1).*cellT.dendriteStd(cellClass) + cellT.dendriteMean(cellClass);
           end
```

III. GOING BEYOND THE DATA

We next *normalize* each column of the data. Normalization before doing PCA is strongly advisable as some variables, say, axon length, are expressed in terms of rather large numbers, whereas some variables, such as spontaneous firing rate, usually take on small values. As we will see, PCA is driven by explaining maximal variance in the data with the minimal number of factors. As variance scales with magnitude, the larger variables will overshadow all others. So we need to express them in the same units, usually z-scores. This is usually done by normalization—taking each value, subtracting off the *mean* value for that variable, and then dividing by the standard deviation of that variable. Alternatively, we could do the PCA on the correlation matrix instead of the covariance matrix—but we are getting ahead of ourselves.

Pseudocode	z-score by column-wise normalization
Python	`sataset = (sataset - sataset.mean())/sataset.std()`

MATLAB	`sataArray = table2array(sataSet);`
	`sataArray = (sataArray - ...`
	`repmat(mean(sataArray),[size(sataArray,1),1]))./ ...`
	`repmat(std(sat`
	`aArray),[size(sataArray,1),1]);`

In Matlab, we convert back to an array because operations like this are easier to do in a matrix. It is not unusual to store data in a struct or a table (as the field or variable names make it easier to remember what the data represent), then convert to a matrix and do the calculations on the array.

In Python, the normalized sataset should look something like Fig. 8.3.

Now we are in a good position to do the PCA.

	Latency	Volume	z	Maxrate	Spont	WaveWidth	Axon	Dendrite
0	-0.183845	1.334715	1.899620	-1.550533	2.569715	-0.773963	2.689569	2.002100
1	-0.296883	-1.206992	-0.686748	0.468447	-0.490398	1.385306	-0.471715	-1.125327
2	-0.464596	-1.094963	0.362676	-0.198085	0.382702	-0.940745	-0.391524	-0.168058
3	0.096252	0.557016	-0.815814	0.859280	-0.428910	-0.938230	-0.394877	0.465161
4	-0.361812	0.061826	-0.844212	0.635258	-0.617164	1.278414	-0.355046	-0.185362

FIGURE 8.3 Example *DataFrame* using *Pandas*. Due to the fact that we always draw randomly from normal distributions, your sata will most certainly look different from what is shown here, but the overall structure should be the same.

Doing PCA involves five basic steps:

1. Calculating the covariance matrix between the variables;
2. Extracting the factors by rotation;
3. Determining the number of factors;
4. Interpreting the meaning of factors;
5. Determining the factor values of the original variables.

Let's go through each step in turn. Before doing that, we do step 0: Initializing and loading in the data (or creating the sata), then normalizing it. As we already did step 0 in the code above, we can proceed directly to step 1.

CALCULATING THE COVARIANCE MATRIX BETWEEN VARIABLES

The point of the PCA is to find axes in a highly dimensional space that express the information contained in the data with less variables than the original dataset. This can work if there are correlations (or redundancies) between variables. If there are substantial correlations, we might be able to find a new variable that is able to express the extent of two (or more) original variables at once. But for this to work, we need to first determine how much variables covary, so we calculate the covariance matrix (Fig. 8.4).

```
covSATA =
      1.00      0.44      0.53     -0.59      0.50     -0.13      0.60      0.18
      0.44      1.00      0.59     -0.33      0.32      0.22      0.57      0.36
      0.53      0.59      1.00     -0.32      0.58      0.07      0.70      0.39
     -0.59     -0.33     -0.32      1.00     -0.12      0.18     -0.39      0.02
      0.50      0.32      0.58     -0.12      1.00     -0.31      0.56      0.28
     -0.13      0.22      0.07      0.18     -0.31      1.00     -0.04      0.16
      0.60      0.57      0.70     -0.39      0.56     -0.04      1.00      0.46
      0.18      0.36      0.39      0.02      0.28      0.16      0.46      1.00
```

FIGURE 8.4 The covariance matrix. As you see, the values on the diagonal are 1—this is a nice manipulation check. A normalized variable covarying with itself should covary perfectly. Note that the matrix is symmetric, as covariance is not directional. Some things covary highly, such as axon length and cortical depth, at 0.7—it might be prudent to try to find a variable that captures this correlational structure and express the information contained in both with a single variable. Others, such as dendrite width and firing rate, seem almost perfectly uncorrelated, whereas there are even some strongly negative correlations, e.g., between latency and maximal firing rate, which can perhaps also be captured by PCA. The absolute values of your covariance matrix will vary, as we draw the sata afresh from normal distributions every time the program is ran. If we wanted to get the exact same values, we would need to seed the random number generator to the same initial value.

Pseudocode	
1. Calculate the covariance matrix	

Python	MATLAB
covSATA=sataset.cov() #1	covSATA = cov(sataArray) %1

FACTOR EXTRACTION AS AN AXIS ROTATION

The point of the PCA is to express the data in terms of a coordinate system with fewer, new axes. But first, we have to find the orientation of these axes in the high-dimensional space. We start with one axis in the direction in which the data vary the most. We can illustrate this in terms of an example with two variables, where we manage to find a single variable that expresses the information contained in the original two variables almost as well, due to the high correlation between the variables. We then subtract this axis from the data and repeat the game with the residuals—residuals orthogonal to the axis we just found. Thus, PCA ends up with a number of axes that are all orthogonal to each other and that account for progressively less variation in the original data.

This might sound, more than just, a bit abstract, so it is best to visualize it (Fig. 8.5).

The program that created this figure is somewhat involved and can be downloaded from the companion website. It makes an animation that shows how the vector is swept around the clock and how the histogram of dot products changes at the same time.

The vector that explains most of the variation between these variables is called the *eigenvector* of the system. The length of this vector is the *eigenvalue*. The first such vector to be extracted is that with the largest eigenvalue (the first principal component), then with the second largest, and so on, until as many factors as input variables (here eight) are determined by PCA.

If you don't understand what eigenvectors are, you are in good company (Gladwell, 2002). Try to imagine a vector that doesn't change direction when a linear transformation is applied to it:

$$\mathbf{L}(\mathbf{v}) = \lambda\mathbf{v}$$
$$\mathbf{A}\mathbf{v} = \lambda\mathbf{v} \tag{8.1}$$

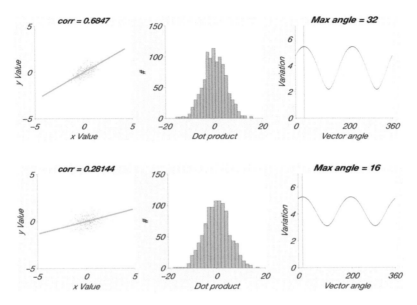

FIGURE 8.5 How PCA works. Top panel: Situation of high correlation. Left panel: A vector is swept around the circle. Blue dots represent cases, and there is a correlation of ~0.68 between the x- and y-dimension. Middle panel: Histogram of dot product between sweeping vector and vectors that represent the dots in the left panel. Right panel: Variance of the histogram in the middle panel, per rotation angle. In this case, the largest variation is at 32 degrees. Bottom panel: Same, but with lower correlation between variables x and y (~0.28). Note that the modulation range between highest and lowest variation has shrunk, compared to the high correlation case. Relatively speaking, the width of the peaks has increased. Of course, the pattern repeats after 180 degrees, as the situation is rotation symmetric.

where $L(v)$ linear transformation of a vector = just a scaled version of the vector—scaled by λ, the eigenvalue. The direction is not changed. This can also be represented by a matrix multiplication, where the matrix A is multiplied to vector v, which doesn't change its direction.

So far the equations. How about a visual? Picture the rotation of the earth. All points that lie on the line that connect the poles lie on the eigenvector of the system (Fig. 8.6).

FIGURE 8.6   Principal axis of rotation of the earth. Points on this hashed line lie on the Eigenvector of the system. They don't change position as the earth spins. Narcissists think they lie on the Eigenvector of every system. Always. In addition, it has to be the one with the largest Eigenvalue. Naturally.

The reason why this factor extraction works is due to the *fundamental theorem*: again, this is an inherently platonic idea—the (messy) data represents shadows cast by (pure) platonic factors, living in the world of ideas and that combine linearly:

$$x_{kj} = a_{j1} \cdot p_{k1} + a_{j2} \cdot p_{k2} + a_{jQ} \cdot p_{kQ}$$
$$X = P \cdot A'$$

(8.2)

EQUATION 8.2   The fundamental theorem. Data x can be expressed in terms of factors p combined linearly and multiplied by weights a.

How can we determine eigenvectors and eigenvalues of a covariance matrix in POM?

Pseudocode
1. Calculate eigenvectors and eigenvalues of the covariance matrix

Python	MATLAB
Eve,Eva=np.linalg.eig(covSATA) #1	[Eve Eva] = eig(covSATA); %1

Eve will contain the eigenvectors (one per column), Eva the corresponding eigenvalues in the diagonal (Fig. 8.7).

Moreover, you will see that the eigensum or eigenmass that is accounted for by the eigenvalues sums to 8, which is a comforting manipulation check that we did the normalization right. In normalized variables, each variable introduces an eigenvalue of 1, as we put in eight variables, the eigensum should be 8. And it is:

Pseudocode
1. Determine the eigensum

Python	MATLAB
eigenSum = np.sum(Eve) #1	eigenSum = sum(Eva(:))%1

FIGURE 8.7   Eve and Eva.

If you look at Eve (or Eva) you will see that eight principal components were extracted. We put in eight variables, so, thus far, we have gained nothing in terms of why we are doing the PCA in the first place, reducing the number of variables we have to worry about. PCA will extract as many principal components as input variables. So we need a proper stopping rule—on to step 3.

DETERMINING THE NUMBER OF FACTORS

As you saw above, PCA will extract as many factors as there are variables. This defeats the point. How do we know how many of these factors are essential, meaningful—in other words when to stop extracting more factors? There are several basic ways to address this issue. The simplest stopping criterion is the "Kaiser criterion" (Yeomans & Golder, 1982). It is fairly simple—only keep factors with eigenvalues larger than 1, the idea being that each variable increases the eigensum of variance that needs to be explained by 1, so if we have to add a factor that explains less than its own addition, it is a liability and should not be included. Due to its sheer simplicity, the Kaiser criterion is still rather popular. The problem is that it is rarely suitable for neuroscience data, as doing PCA on a large multivariate dataset might yield 20 or 30 factors with eigenvalues above 1. That's still a lot of factors to keep track of, so it doesn't help much. A different approach consists in keeping as many factors as are needed to explain 90% of the variation in the dataset. It is of limited use in neuroscience for similar reasons—to account for 90% of the variation in a large multivariate neural dataset, one will need to retain an unreasonably large number of factors, defeating the purpose of doing a PCA in the first place.

Another popular approach is to arrange the factors by magnitude (this is called a "Scree plot") and look for an "elbow." The factors left of the elbow are considered to be meaningful, whereas the others are usually thought to correspond to noise. The problem with this approach is that it is inherently subjective, relies on visual inspection, and doesn't appear too principled. Also, there might not be an obvious elbow in the data.

Here is the code to create a scree plot for this sataset in POM (Fig. 8.8).

Pseudocode	
	1. import PCA package
	2. create PCA model
	3. set the percent of data on which to train the model, use 100 for this example
	4. calculate what index of the sataset that percent corresponds to
	5. fit the PCA transform to the training set
	6. set the y-training data to the celltype
	7. extract just the y-training values (not the indices)
	8. plot the explained variance as a function of principal component
	9. plot a horizontal line at 1
	10. set the title, xlabel, and ylabel

(Continued)

| Python | ```
from sklearn.decomposition import PCA #1
pca = PCA() #2
trainPercent = 100 #3
trainNum = int(len(sataset)*trainPercent/100.) #4
xTrain = pca.fit_transform(sataset[:trainNum]) #5
yTrain = celltypes[:trainNum] #6
yTrain = [_[0] for _ in np.array(yTrain)] #7
plt.plot(pca.explained_variance_ratio_*8) #8
plt.axhline(1,c='k',ls='--') #9
plt.title('Screeplot of Sata for 100 simulated neurons') #10
plt.ylabel('Eigenvalue') #10
plt.xlabel('Principal Component #') #10
``` |
|---|---|
| MATLAB | ```
eigVal = sortrows(sum(Eva)',-1);
plot(1:length(eigVal),eigVal,'color','b','linewidth',3);
line([1 length(eigVal)], [1 1],'color','k','linestyle','--')
xlabel('Principal component #')
ylabel('Eigenvalue')
title('Screeplot of Sata for 100 simulated neurons')
``` |

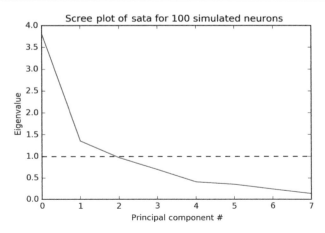

FIGURE 8.8   Scree plot of sata principal components. Blue: eigenvalues in order of decreasing magnitude. Black dashed line: Kaiser criterion. Both Kaiser and Elbow criteria seem to agree in this case, suggesting two valid principal components.

How about the 90% of variance explained criterion? (Fig. 8.9).

Pseudocode	1. plot the cumulative explained variance as a function of principal component 2. draw dashed line at y = 0.9 3. set the title, xlabel, and ylabel
Python	``` plt.plot(np.cumsum(pca.explained_variance_ratio_)) #1 plt.axhline(.9,c='k',ls='--') #2 plt.title('Screeplot of Cumulative Variance Explained') #3 plt.ylabel('Cumulative variance explained') plt.xlabel('Principal Component #') ```

MATLAB	```
	cumEigVal = cumsum(eigVal)./sum(eigVal); %1
	plot(1:length(cumEigVal),cumEigVal,'color','b','linewidth',3);
	line([1 length(eigVal)],
	[0.9 0.9],'color','k','linestyle',':') %2
	xlabel('Number of principal components used')%3
	ylabel('Cumulative variance explained')
	title('Screeplot of cumulative variance explained')
	``

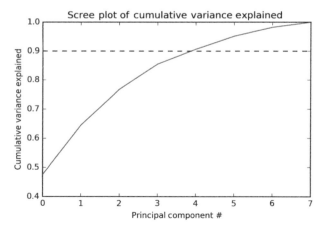

FIGURE 8.9 Cumulative variance explained by increasing numbers of principal components. As you can see, we would need five principal components to account for more than 90% of the variance in the dataset. A principled, but rather liberal (and thus less useful) criterion for typical neural data.

Thus, we'll recommend to use (and implement) one more method to determine the number of principal components, as it is perhaps particularly suitable in neuroscience.

This final method we discuss here is Horn's "parallel analysis" approach (Horn, 1965; Ledesma & Valero-Mora, 2007). This supports the notion that 1965 was a seminal year in the development of computational approaches—the same year Bracewell published his book on the Fourier transform and Tukey developed the FFT. The basic idea is that one does a PCA on an array of random numbers of the same dimensionality as one's real data. Doing so, one can get an estimate of the magnitude of the eigenvalues that a PCA of pure noise would yield. One could reasonably expect real factors to exceed this magnitude. In order to get an estimate of how variable these noise estimates are, one should do this lots of times, so a basic drawback of Horn's method is that it is computationally by far the most "expensive" of all these frameworks, but that is not a big drawback in the age of fast and readily available computing power. A bigger issue is which noise distribution to use, from which to draw. Horn himself used a normal distribution, and this is fairly standard, but distributions of many parameters in neuroscience as well as the social sciences do not follow normal distributions and are better described by power laws or the like (Fig. 8.10).

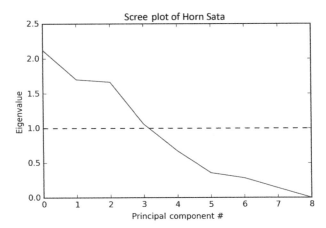

FIGURE 8.10   Scree plot of the Horn Sata. This is a plot of a single horn trial—just one matrix filled with random noise drawn from a normal distribution. Note that the magnitude of principal components is not zero, but substantially less than for real data. We can use this to get an idea of what one could reasonably expect purely by chance.

Pseudocode	Python
1. create 8x100 array of random numbers	`hornmat = np.array(np.random.randn(8,100)) #1`
2. create PCA model	`pca = PCA() #2`
3. selected percent of the sata to use (use this value for smaller traning sets, becomes relevant in Ch. 9)	`trainPercent = 100 #3`
4. calculate the integer index for training	`trainNum = int(len(hornmat)*trainPercent/100.) #4`
5. transform the sata via PCA	`xTrain = pca.fit_transform(hornmat[:trainNum]) #5`
6. extract the y-training sata	`yTrain = yTrain[:trainNum] #6`
7. get just the values (not the indices)	`yTrain = [_ for _ in np.array(yTrain)] #7`
8. plot the explained variance for each PC	`plt.plot(pca.explained_variance_ratio_*8) #8`
9. plot a horizontal line at y=1	`plt.axhline(1) #9`
10. set the title	`plt.title('Screeplot of Horn Sata') #10`

Now, let's simulate over a thousand trials to get an idea what the noise distribution looks like (Fig. 8.11).

Pseudocode
```
 1. create figure
 2. create subplots
 3. declare empty lists to hold the first PC of each simulation
 4. for 1000 simulations of the horn sata
 5. create 8x100 array of random numbers
 6. create PCA model
 7. selected percent of the sata to use (use this value for smaller traning sets, becomes relevant
    in Ch. 9)
 8. calculate the integer index for training
 9. transform the sata via PCA
10. extract the y-training sata
11. get just the values (not the indices)
12. calculate the amount of explained variance for each PC
13. append the variance explained by the first PC to the list
14. plot the explained variance for each PC
15. plot the cumulative explained variance for each PC
16. plot a horizontal line at y=1
17. calculate the 97.5% confidence (as the 975th entry in 1000-length sataset)
18. plot a dashed line at the horn confidence line
19. set the title
20. set the labels
21. set the title
```

(Continued)

Python

```
fig = plt.figure(figsize=(22,10)) #1
ax = plt.subplot(121);ax2 = plt.subplot(122)#2
firstPC = [] #3
for hornNum in range(1000): #4
    hornmat = np.array(np.random.randn(8,100))#5
    pca = PCA() #6
    trainPercent = 100 #7
    trainNum = int(len(hornmat)*trainPercent/100.) #8
    xTrain = pca.fit_transform(hornmat[:trainNum]) # 9
    yTrain = celltypes[:trainNum] # 10
    yTrain = [_[0] for _ in np.array(yTrain)] #11
    explained = pca.explained_variance_ratio_*8 #12
    ax.plot(explained,alpha=0.5,c='k') #14
    ax.plot(explained,alpha=0.5,c='k') #14
    ax2.plot(np.cumsum(pca.explained_variance_ratio_),alpha=0.5,c='k') #15
ax.axhline(1) #16
hornconf=sorted(firstPC)[975] #17
ax.axhline(hornconf) #18
ax.set_title('Screeplot Horn Sata, 97.5 confidence interval = '+str(hornconf)[:4]) #19
ax.set_xlabel('Principal Component #') #20
ax.set_ylabel('Eigenvalue') #20
ax2.set_title('Screeplot of Cumulative Horn Sata') #20
ax2.set_xlabel('Principal Component #') #20
ax2.set_ylabel('Cumulative variance explained') #20
plt.savefig('Screeplot of Horn Sata 1000 trials.png',dpi=300) #21
```

In the interest of saving space (and to introduce the MATLAB file exchange) we didn't show the corresponding MATLAB code here. Instead, a great and well documented implementation of this algorithm in MATLAB by Lanya Cai can be downloaded from https://goo.gl/8IFGZ0—the function in question is called HornParallelAnalysis and is quite efficient.

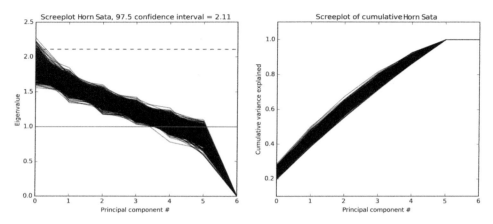

FIGURE 8.11   Simulation of Horn Sata plot **1000×** . Drawing a horizontal line at the first principal component of the 975th ordered eigenvalue gives us an upper bound of a 95% confidence interval[96] for accepting principal components. We can now overlay the real scree plot with the distribution that was simulated by noise and determine which factors escape the 95% confidence interval around the noise distribution—and we can consider those as "real" and not due to random noise.

INTERPRETING THE MEANING OF FACTORS

Now that we have found the principal components (factors) and determined how many we consider to be meaningful, we need to figure out what they actually mean. Interpreting the meaning of factors is notoriously difficult and controversial. It is usually done by looking at how clusters of variables load on the underlying factors and using their known meaning to determine some commonalities. This is how the factor names for the personality psychology example above were determined, but it can be hard to arrive at a consensus. For instance, conscientiousness is statistically almost indistinguishable from "grit," so how would we know whether to call the factor grit or conscientiousness (Rimfeld, Kovas, Dale, & Plomin, 2016)? In neuroscience, the problem can be even worse. It is usually not obvious what to call a handful of factors that can account for most of the behavior in a neural network in state space (Mante, Sussillo, Shenoy, & Newsome, 2013; Kaufman, Churchland, Ryu, & Shenoy, 2014). However, for the sake of behavioral prediction or classification, this is usually not even necessary. For instance, spike sorting methods routinely rely on "cluster cutting" in PCA space where the meaning of the underlying dimensions is not always clear and only loosely corresponds to anything obvious like spike amplitude or spike width. As the point of the exercise is simply to classify which neural cluster a given spike most likely came from, that might be fine. It is not fine if you are in the process of a grand theory of how the brain works that relies on PCA—in that case, it might be necessary to have at least a hypothesis of what the underlying factors represent. This step is entirely up to the ingenuity of the researcher.

As far as we can tell, the two meaningful factors we extracted to correspond to our sata are "anatomical aspects" versus "physiological aspects" of the neurons.

DETERMINING THE FACTOR VALUES OF THE ORIGINAL VARIABLES

What remains to be done? To express the original data in terms of the factors we found. To do so, we revisit the fundamental theorem (Eq. 8.2), but solve for P:

$$P = X \cdot (A')^{-1} \tag{8.3}$$

EQUATION 8.3   The fundamental theorem, solved for P.

Let's do this manually, in POM.

Pseudocode

1. Flip the eigenvectors so they are sorted from largest to smallest
2. Calculate rotated values
3. Open new figure
4. Plot rotated values in 2D space
5. Add axes labels
6. Format plot to make it look square

Python	MATLAB

```
rotatedSataArray = np.dot(np.array(sataset),Eve); #2
plt.figure() #3
plt.scatter(rotatedSataArray[:,0],rotatedSataAr
ray[:,1]) #4
plt.xlabel('PCA 1'); plt.ylabel('PCA 2') #5
plt.axis('equal') #6
```

```
flipEve = fliplr(Eve); %1
rotatedSataArray = sataArray * flipEve; %2
figure %3
plot(rotatedSataArray(:,1),rotatedSataArray(:,2),'o') %4
xlabel('PCA 1'); ylabel('PCA 2') %5
axis equal; axis square %6
```

Once we have calculated the values of the 100 simulated neurons in terms of the two principal components we deem real, we can just plot them—a 2D plot is quite feasible and lends itself to visual inspection (Fig. 8.12).

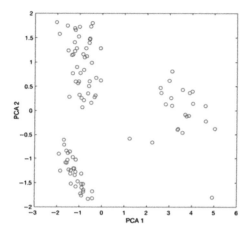

FIGURE 8.12 Plotting our simulated neurons in terms of the rotated values, PCA1 versus PCA2.

That was a lot of work! Lots of steps, lots of moving parts, lots of things to keep track of. The good news is that now that you, hopefully, understand what is going on, you'll probably never again do this step by step, by hand. Instead, there are functions in POM that do all of this at once.

Pseudocode	1. Determine principal components all in one step
Python	`pca = PCA(n_components = 2) #1`
MATLAB	`[Eve, rotVal, Eva] = princomp(sataArray); %1`

We can now plot the results of *that*, all in one fell swoop, and label it while we are at it (Fig. 8.13).

Pseudocode	1. create figure 2. create marker shapes 3. create marker colors 4. create marker labels 5. for each value in the y-training set 6. scatter plot the data so each value in y-training is colored by group 7. label axes 8. set title 9. show the legend 10. make axes square

Python

```
fig=plt.figure(figsize=(6,5)) #1
markershape = ['o','^'] # circle and triangle #2
markercolor = ['red','cyan'] #3
markerlabel = ['inh','exc'] #4
yTrain = np.array(celltypes['transmission'])
for yind,yval in enumerate(set(yTrain)): #5

    plt.scatter(rotatedSataArray[yTrain==yval,0],\
                rotatedSataArray[yTrain==yval,1],\
                c=markercolor[yind],marker=markershape[yind],label=markerlabel[yind],s=50) #6
plt.xlabel('PC1') #7
plt.ylabel('PC2') #7
plt.title('First two principal components of sata') #8
plt.legend() #9
plt.axis('equal')#10
```

MATLAB

```
figure
toPlot1 = find(cellTypes.transmission == 0);
toPlot2 = find(cellTypes.transmission == 1);
h1 = plot(rotatedSataArray(toPlot1,1),rotatedSataArray(toPlot1,2), ...
    'o','markerfacecolor','r', 'markeredgecolor','k')
hold on
h2 = plot(rotatedSataArray(toPlot2,1),rotatedSataArray(toPlot2,2), ...
    '^','markerfacecolor','c','markeredgecolor','k')
title('First two principal components of sata')
legend([h1 h2], {'inh','exc'})
xlabel('PCA 1'); ylabel('PCA 2')
axis equal; axis square
set(gcf,'color','w')
```

III. GOING BEYOND THE DATA

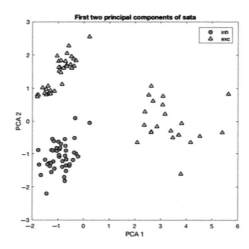

FIGURE 8.13   First two principal components of the *sata*, where each cell type is labeled according to its type of *excitatory (exc)* or *inhibitory (inh)*.

   Now that we are able to represent these neurons in a low-dimensional space that is easy to visualize, let's try to see whether we can classify them in the Chapter 9, Classification and Clustering.

   We have reached a reasonable point to call "1"—a good intuition for what a PCA is, why we do it, and how to do one. What lies beyond 1? Other and more advanced dimensionality reduction methods include factor analysis and independent components analysis (ICA). All of these extend the PCA in some way, e.g., PCA assumes Gaussian components and forces orthogonal principal component. Relaxing these constraints can yield better factor loadings (Comon, 1994). For instance, the ICA allows to distinguish sources (think sound sources at a cocktail party) that might not be orthogonal—it simply identifies factors that are statistically independent (usually nonnormal). One drawback of ICA is that while the order of the factors is well defined in PCA (it starts with the one that yields the highest eigenvalue, then the second highest orthogonal one and so on), order is somewhat arbitrary in ICA, as it depends on the starting position—doing this repeatedly might yield different results. It's usually a bad sign if a result is not robust against multiple attempts at implementing the same method. The development of dimensionality reduction methods is an active area of research, particularly the development of methods that can be suitably applied to neuroscience use cases such as minimum average variance estimation (MAVE) by Xia, Tong, Li, & Zhu (2002).

⌘. *Pensee:* How much confidence should we place in confidence intervals? It depends—confidence intervals as a concept were controversial from the start (Salsburg, 2001)- but perhaps quite a bit more than in *p*-values. Here is why. A confidence interval is an interval estimate of a population parameter. It specifies the expected probability that a given interval will contain the population parameter of interest, usually the mean (Neyman, 1937). For instance, if we, hypothetically, repeated a study 1000 times with a 95% confidence interval, the mean would fall within the confidence interval 950 times. This is a reasonable attempt to remedy a problem that has bedeviled much of science as of late. Under pressure to publish "positive" results, scientists exploit "researcher degrees of freedom" to publish false-positives (Simons, Nelson, & Simonsohn, 2011). This is called "p-hacking" and has led to an avalanche of papers with p just below 0.05, but that yield little to no evidentiary support in favor of the hypothesis they were supposed to test (Vadillo, Hardwicke, & Shanks, 2016). Arguably, such studies are worse than useless as they are misleading instead of just agnostic in terms of the position of a field on the status of an effect that does not really exist (OSC, 2015). In contrast, narrow confidence intervals can provide estimates of how large an effect is. Even if there is no effect, a narrow confidence interval can be useful ("evidence of absence"), as opposed to the wide confidence intervals that characterize underpowered studies ("absence of evidence"). How can confidence intervals help? Quite a bit, as they contain much more information than a *p*-value (see Fig. 8.14).

FIGURE 8.14 Six confidence intervals. Say we want to know whether a drug like NZT 48 improves IQ. To test whether there is an effect, we give NZT 48 to an experimental group, but not a control group, and test the difference to see whether the group means differ in IQ afterwards. The blue line represents zero difference between the two groups. The six vertical bars correspond to different hypothetical confidence intervals (arbitrary units) that represent potential outcomes of the experiment. The leftmost CI represents a tricky situation—as the CI contains 0, the difference in sample means won't be significant. But we are unsure about this estimate—the effect could be considerable, as the upper edge of the CI is quite high up, but it includes the possibility of a zero effect more often than we are comfortable with. The CI immediately to the right of that also contains zero, but is also associated with considerable uncertainty—NZT 48 might even hurt IQ, if values at the lower edge of the CI are in fact real. We could paraphrase this as absence of evidence—we don't really know what is going on with NZT 48. The CI immediately to the right of that represents pretty compelling evidence that there is really no difference, and as the CI hugs the zero line closely, we can be quite confident in that. Thus, this one represents evidence of absence (of an effect). A *p*-value lumps all three of these situations together and will yield $p > 0.05$, not significant. But it really does matter which of these situations is the case—being reasonably certain that there is really no effect can be quite valuable (e.g., when ruling out "side effects" (there

FIGURE 8.14   (*Continued*)

are really only treatment effects, some of which are undesirable and which we call "side effects," but their magnitude can actually dwarf the "main effects," whereas "side" can connote "negligible") of a medication). In contrast, the three CIs on the right all correspond to various outcomes that a *p*-value would lump together as $p < 0.05^*$. But like in the nonsignificant cases, it really does matter which situation is the case. The leftmost one is usually a tell-tale sign of a *p*-hacked study. A giant confidence interval that just barely excludes zero. Plus, we have no idea whether the effect is almost negligible or considerable. The CI to the right of that is somewhat better (we are comfortably away from zero) so *p*-hacking is less likely and a sizable effect is more likely, but we don't know quite how sizable—could be gigantic, could be rather modest. From the perspective of the experimenter, the CI on the far right is the best outcome—there is a high likelihood of a strong effect of NZT 48 on IQ (we sure could use it) and we can be quite certain or confident, of its magnitude.

CHAPTER

9

Classification and Clustering

When analyzing the neural data in Chapter 4, Correlating Spike Trains, we were struck by the diversity of neural response profiles, which is fairly typical for data resulting from electrophysiological recordings. This raises the obvious question whether we are even recording from the same type of neuron on all electrodes, or whether there are distinct groups of neurons which should be analyzed separately. For instance, recall that the response profiles of some neurons were strongly correlated, whereas those of others were strongly anticorrelated. As the responses were elicited by the same stimuli, one can reasonably wonder whether one group of neurons corresponds to glutaminergic pyramidal cells, whereas the other could correspond to GABAergic inhibitory interneurons. Lumping them all together might be misleading, yielding middling degrees of intercorrelation and perhaps prevent a deeper understanding of the local neural circuitry. Averaging is a radical act—throwing away most of the information in order to arrive at parameters such as the sample mean that is presumably rather stable (Stigler, 2016). This works reasonably well if the things one averages over are all members of the same set and the observed variation is entirely due to noise. If they are not, one can get in trouble. For instance, some people have a mutation that prevents them from metabolizing folic acid to an active form important in methylation (Weisberg, 1998). Simply giving a drug that increases methylation to everyone might mask the beneficial effects of the drug for the subgroup with the mutation, as many people (without the mutation) will suffer effects of overmethylation. If the two effects balance in the population this might even seemingly produce a non-result, as the overall mean doesn't change. So it can be useful to look for clusters of response patterns in data before averaging, as they might represent structurally or functionally distinct subgroups. But it is important that this is done in

Neural Data Science.
DOI: http://dx.doi.org/10.1016/B978-0-12-804043-0.00009-X

a principled fashion, as humans are prone to seeing meaningful patterns in random noise (Shermer, 2008; Fyfe, 2008; Liu et al., 2014).

Data from extracellular recordings are often limited in terms of addressing what kinds of cell class one recorded from, although there have been attempts to infer cell type from the shape of the waveform (Mitchell, Sundberg, & Reynolds, 2007). Thus, for educational purposes, we will build on the results from Chapter 8, Dimensionality Reduction, as the starting point of our classification attempts here, as we know ground truth. The goal of our exercise is to see whether we can identify distinct cell classes in our sataset.

Overall, the sample mean of peak firing rate in our sataset is 0.67 sp/s with a 95% confidence interval of ±0.06 sp/s (your values are likely to be a bit different, as they are generated randomly and we deliberately neglected to provide a seed in order to emphasize the fact that sata is created, usually from randomness). That is fine, but not very meaningful—the sample mean is supposed to be an estimator of the population mean (of cells in general, not just *these* cells) and the population mean the *expected value* of what one could reasonably expect if one was to pick a cell at random. If the confidence interval is narrow, this is no problem. If it is large (as in this case) the utility of the mean estimate is rather limited. To see why, imagine you are tasked with designing plane seats in the most economical fashion but yet comfortable enough that people will want to fly with your plane. A key design constraint will be how tall the person sitting in it can be expected to be. If you are drawing from a homogeneous population, this is not much of a problem. Heights will likely be distributed normally, with most people clustering narrowly around the mean. But imagine working for an airline of a virtual country that is half populated by NBA players (average height about 200 cm or 6′7″) and half by pygmies (average height about 150 cm or 4′11″). Assuming that both subgroups are equally likely to fly, you face quite the conundrum—the height distribution will be *bimodal*, with a considerable separation between the peaks (assuming a standard deviation of 8–10 cm). In other words, one size (of seat) will most definitely not fit all. Taking these subpopulations into account when designing seats is critical. This might sound like a silly example, but seats of fighter jets have to be customized to the individual pilot. Averaging won't do - pilots are too different in too many dimensions, and plane crashes have been attributed to averaging inappropriately in this way (Rose, 2016). Translating this back to our analysis of neural firing rates: Can we refine this analysis by identifying subsets of neuron classes first, then describing group sample means that better describe the spiking characteristics of that cell type—perhaps yielding tighter confidence intervals of these new means?

A first inkling that this might be necessary can be gleaned from a histogram of maximal firing rates in the sataset in POM (Fig. 9.1).

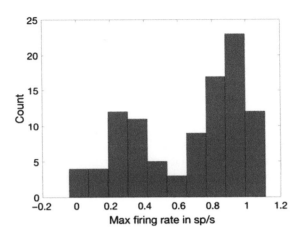

FIGURE 9.1 There seems to be two subpopulations of neurons: one that fires maximally around 0.3 sp/s and one that fires around 1 sp/s.

Pseudocode
1. Open figure
2. Make a histogram of the max firing rates of the sataset
3. Add axes labels

Python	MATLAB
f = plt.figure() #1	figure %1
plt.hist(sataset['maxrate']) #2	hist(sataset.maxrate) %2
plt.xlabel('Max firing rate in sp/s') #3	xlabel('Max firing rate in sp/s') %3
plt.ylabel('Count')#3	ylabel('Count') %3

This looks like a bimodal distribution, which strongly hints at several different underlying populations of neuron types. But such a visualization is not enough to make the case that there is a statistically reliable way of classifying these cells in subgroups. Then again, if subgroups exist, it is important to differentiate their behavior—lumping them all together will yield misleading results. Thus, the question whether such subpopulations exist and how to distinguish them (e.g., anatomically, physiologically, or functionally) is often the subject of intense scientific scrutiny, e.g., the distinction between simple and complex cells in V1 (Hubel & Wiesel, 1968; Mechler & Ringach, 2002) and pattern and component cells in MT (Movshon, Adelson, Gizzi, & Newsome, 1992; Jazayeri, Wallisch, & Movshon, 2012).

Let's start with a classification method that we already encountered in Chapter 7, Regression—logistic regression. Here, we apply it to the question whether we can reliably identify/separate subgroups of neurons that we suspect to have different properties, such as excitatory versus inhibitory cells.

Classification is useful when we can make predictions about data that we haven't seen yet. In this case, we have access to all of our sata, but we can set some (say, 10%) of it aside. This leaves us with 90% of the original sataset. We will create a model to classify groups based on properties of the cells using this 90%. The reason we set the 10% aside is that we strongly discourage using the same sata or data one uses to *build* the model to also *test* it. That's double-dipping and a surefire way to create a model that will be overfitting. In other words, we use the 90% of the sata to build the model and the remaining 10% to test it and see whether the classification performance generalizes. In machine learning terms, the 90% of the data here are known as our *training data*, where the *x-training data* are all of the properties of the cells, and the *y-training data* are which group the cells belong to. We then take the 10% of the data that we didn't use to build the model as our *testing data*, where we use the properties of these 10% as our *x-testing data*, and then make prediction of what groups these testing cells belong to. We'll compare these predictions to the *y-testing data* to see how well our predictive model performs on data that were not involved in building it. This process is called cross-validation and should always be carried out. If the 90/10 split seems arbitrary, it is. There are many rules and heuristics to guide a proper split. Generally speaking, one wants to use as much data to build the model as possible, but also as much as needed for a fair test. A popular solution to this is the "leave one out" method, where a model is built on *all* the data except for one data point, where we use almost all of the data to predict that one point. Then, we move on and use all the data except for another point and so on, then blend all of these models. The obvious drawback of this is that it is computationally rather expensive, as one has to build n models, if one has n datapoints. Also, if there are interactive patterns that depend on multiple data points, one will obviously not find them by leaving just one out.

We additionally make use of the function meshgrid, which gives us the coordinates of all x–y points in a two-dimensional space. Note that x and y for the meshgrid will correspond to the values of the first and second principal components (PC1, PC2), respectively. The idea is that we take all possible values that PC1 could be (x) and all possible values for what PC2 could be (y), to create a grid of points in a two-dimensional x–y space. Meshgrid gives us these points. We'll be testing how each of the points in this space gets classified. Meshgrid is a popular function that underlies much 3D visualization of functions, so it is important to understand it well before using it in this way.

What meshgrid is really doing is creating two ramps or gradients, one in the x-direction (while keeping y constant) and one in the y-direction while keeping x constant. This is abstract, so it's best to just illustrate that (Fig. 9.2).

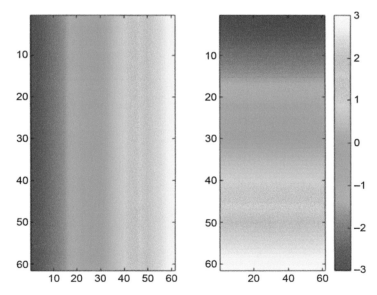

FIGURE 9.2 Visualization of meshgrid—it creates complementary ramps.

Pseudocode

---

1. Create a meshgrid from -3 to 3 in steps of 0.1, for 61 steps
2. Open a new figure and subplot
3. Visualize the points in the matrix X as colors, scaled from min to max, spanning the color space
4. Open a new subplot
5. Visualize the points in the matrix Y in the same way
6. Add a color bar for scale, to see which color corresponds to which value

---

Python	MATLAB
| ```
X,Y = np.meshgrid(np.arange(-3,3,.1), np.arange(-3,3,.1)); #1
f=plt.figure #2
plt.subplot(1,2,1) #2
plt.imshow(X) #3
plt.subplot(1,2,2) #4
plt.imshow(Y) #5
plt.colorbar() #6
``` | ```
[X,Y] = meshgrid(-3:0.1:3 -3:0.1:3); %1
figure %2
subplot(1,2,1) %2
imagesc(X) %3
subplot(1,2,2) %4
imagesc(Y) %5
colorbar %6
``` |

Seems like an obscure thing to do. Why is this useful? Because we can now visualize values in three dimensions, e.g., if an outcome z depends on two inputs x and y, such as in Fig. 9.3.

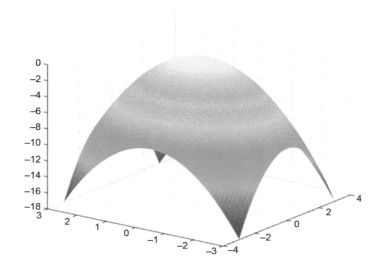

FIGURE 9.3 Nice 3D plot.

Pseudocode

1. Import package for 3D plotting
2. Compute Z as a combination of the squared values of x and y, added, at each point in the matrix, creating a parabola opened downward by multiplying with -1
3. Open a new figure
4. Image the X, Y and Z values as a 3D surface
5. Take grid off
6. Get a decent view

Python	MATLAB
| ```
from mpl_toolkits.mplot3d import Axes3D #0
Z = -1.*(X**2 + Y**2); #1
fig = plt.figure() #2
ax = fig.add_subplot(111, projection='3d')
h = ax.plot_surface(X,Y,Z) #3
``` | ```
Z = -1.*(X.^2 + Y.^2); %1
figure %2
h = surf(X,Y,Z) %3
set(h,'linestyle','none') %4
set(gca,'View',[-54 24]) %5
``` |

On to putting it all together—including utilizing what we learned in Chapter 7, Regression, and the sata from Chapter 8, Dimensionality Reduction. For instance, we'll use the meshgrid to basically go through all x-values while keeping the y-value constant (scanning a line), then doing the same again for the next line of x-values until we "rastered" the entire sata space (Fig. 9.4).

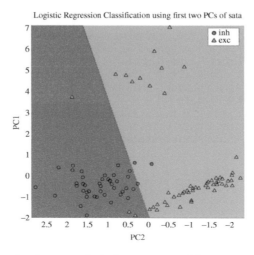

FIGURE 9.4 Logistic regression classification using first two principal components of the sata.

Pseudocode

1. import LogisticRegression package
2. import ListedColormap package
3. Create a logistic regression model called logRegression
4. Fit the model to the variables xTrain and yTrain (from Chapter 8)
5. Create figure
6. Specify marker shapes: circle and triangle
7. Specify marker colors red and cyan
8. Specify marker labels for *inhibitory* and *excitatory* cells
9. Specify grid resolution
10. Create a colormap based on the markercolor variable
11. Calculate the absolute maximum value of the xTrain array. With a numpy array you use the function array.max() to get its maximum. We also allow the negative of the minimum (determined by array.min()) to be considered as the maximum. This absolute maximum will be used as the outer limits of the space in which we test what neurons belong in which class. Note that we add 0.1 to this value to give us a little more room in the plot—with a +0.1 pad added, no point in the sataset will be plotted on the edge of our visualization
12. Set the minimum plotting parameter to the negative of the max—this is so that the plot minimum axis is equal in distance from zero as the plot maximum
13. Create a meshgrid that covers all x and y values (just like above). This meshgrid allows us to predict what group that any range of data would belong to
14. Predict what the logistic regression model thinks the values for each x and y in the meshgrid should be. This prediction is the core of machine learning: we fit a model using a training data, and then predict what groups a whole range of other points belong to. This is why we use the meshgrid, so that we have a range of space spanning all dimensions of the data to test in which groups data with that space belong
15. The variable Z is now a 1-D array. We need to reshape it so that it is the shape of the xgrid in order to create a contour plot. This is a bit wonky—having to flatten and then reshape—but is just necessary data wrangling so that we don't break any of the functions
16. Plot the meshgrid as a contour, and specify the colormap as determined by line 10
17. For each value in the yTrain
18. Scatterplot the groups with the parameters specified by the marker values above. This clever bit of code lets us plot each of the two groups of sata as different colors (red and cyan for inhibitory and excitatory)
19. Set the x and y labels
20. Set the title
21. Show the legend

Python

```
from sklearn.linear_model import LogisticRegression #1
from matplotlib.colors import ListedColormap #2
logRegression = LogisticRegression() #3
logRegression.fit(rotatedSataArray[:,:2],yTrain) #4
fig=plt.figure(figsize=(6,5)) #5
markershape = ['o','^'] #6
markercolor = ['red','cyan'] #7
markerlabel = ['inh','exc'] #8
gridRes=.02 #9

xTrain = np.array(rotatedSataArray[:,:2])
cmap = ListedColormap(markercolor) #10
maxval = max(-1*xTrain.min(),xTrain.max())+.1 #11
minval = -maxval #12
xgrid,ygrid=np.meshgrid(np.arange(minval,maxval,gridRes),np.arange(minval,maxval,gridRes)) #13
Z = logRegression.predict(np.array([xgrid.flatten(),ygrid.flatten()]).T) #14
Z = Z.reshape(xgrid.shape) #15
plt.contourf(xgrid,ygrid,Z,alpha=0.4,cmap=cmap) #16

for yind,yval in enumerate(set(yTrain)): #17
    plt.scatter(x=xTrain[yTrain==yval,0], y=xTrain[yTrain==yval,1], c=markercolor[yind],\
    marker=markershape[yind], label=markerlabel[yind],s=50) #18
plt.xlabel('PC1');plt.ylabel('PC2') #19
plt.title('Logistic Regression Classification using first two PCs of sata') #20
plt.legend() #21

xTrain = [rotatedSataArray(toPlot1,1),rotatedSataArray(toPlot1,2);...
rotatedSataArray(toPlot2,1),rotatedSataArray(toPlot2,2)];
yTrain= [zeros(length(toPlot1),1);ones(length(toPlot2),1)];
b = glmfit(xTrain, yTrain, 'binomial', 'link', 'logit'); %3
gridRes = 0.02; %9
xMin = min(xTrain(:,1))-0.1; %11
xMax = max(xTrain(:,1))+0.1; %11
yMin = min(xTrain(:,2))-0.1; %11
yMax = max(xTrain(:,2))+0.1; %11
[xGrid,yGrid] = meshgrid(xMin:gridRes:xMax,...
                yMin:gridRes:yMax); %13
```

(Continued)

III. GOING BEYOND THE DATA

MATLAB
```
xFlat = xGrid(:);
yFlat = yGrid(:);
yfit = glmval(b,[xFlat yFlat],'logit'); %14
yfitSurf = reshape(round(yfit),size(xGrid,1),size(xGrid,2)); %15
figure %5
h = surf(xGrid,yGrid,yfitSurf); %16
view([-90 90]);
h.LineStyle = 'none';
xlim([xMin xMax]); ylim([yMin yMax]) %12
hold on
h1 = plot3(rotatedSataArray(toPlot1,1),rotatedSataArray(toPlot1,2), ... %18
2*ones(length(toPlot1)),'o','markerfacecolor','r', 'markeredgecolor','k');
h2 = plot3(rotatedSataArray(toPlot2,1),rotatedSataArray(toPlot2,2), ... %18
2*ones(length(toPlot2)),'^','markerfacecolor','c','markeredgecolor','k');
legend([h1(1) h2(1)], {'inh','exc'}) %21
xlabel('PCA 1'); ylabel('PCA 2') %19
cMap = [1 0 0; 0 1 1]; colormap(cMap); %10
title('Logistic Regression Classification using first two PCs of sata') %20
```

Logistic regression seems to be doing a pretty good job! For this sataset, we might reasonably be satisfied with just the logistic regression classification method. The logistic regression, when used as a classifier with two groups, will assign each point to one of the two groups. In earlier chapters, we also used the logistic regression to predict the probability of outcomes (e.g., in chapter 7: Regression, the probability of getting into grad school was analyzed). While we could have predicted the probability of being in one or another group here, we explicitly wanted to pick which of the two groups the points are more likely to belong to. The outcome of the logistic regression here is a one-dimensional linear classifier, meaning the predicted groups for the input sata will be divided by a straight line. However, it is possible that data may not be cleanly separated by a straight line. We explore other methods to see how such a nonlinear separation of groups might occur.

Another popular method for classifying data into categories is called a "support vector machine" (SVM). That surely sounds more than just a bit daunting—but the idea is actually quite straightforward. A support vector machine is a "widest margin classifier" or even a "maximum margin classifier." These classifiers find the separating line (in higher-dimensional spaces *hyperplane*) that is equidistant to the two classes of datapoints, such that it maximizes the distance between the hyperplane and the nearest datapoint on each side (Boser, Guyon, & Vapnik, 1992). The notion is that a line is selected to evenly split the data into groups, where the line is equidistant to only the closest points in each group. In a logistic regression, adding points far from the decision boundary would influence the slope of the decision line. In contrast, SVM implies that only what happens near the boundary matters, points far from the decision line, have no influence on the orientation of the dividing line. So by looking for the widest margin that separates clouds of datapoints, the SVM is flexible with regard to measurement errors—it needs the widest possible buffer for accurate classification in the presence of noise. The eponymous support vectors after which the method is named are the ones that constrain the line—defined by points at the boundary (see Fig. 9.5).

The separator yielded by SVM does not have to be a straight line, such as the line which separates the two groups in the logistic regression above. Rather, the SVM boundary can be a curved line, or some polynomial. Figuring out which kind of line to use to separate groups may require deeper understanding of the data, or at least require visual inspection of the output. The particular line drawn in the sand between the groups is known here as our *kernel*, similar to the kernels used in Chapter 4, Correlating Spike Trains. In the simplest case of a linear classifier this kernel is just a straight line that separates groups. The SVM can be implemented with a much fancier boundary separating the groups: the kernel can be a customized line, a polynomial, a circle, or more. In the cartoon in Fig. 9.5, we show toy examples of different boundaries we can draw between two groups.

The division of groups by some line that splits them is actually division by a *hyperplane*. A hyperplane is any *plane* that is geometrically one dimension less than the number of dimensions we are interested in dividing. In the simple example

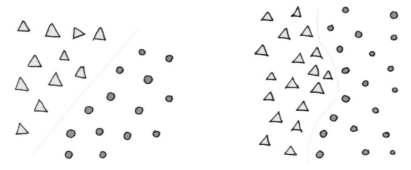

FIGURE 9.5 The line that separates the two groups in an SVM does so to equalize the distance between the nearest points in either group.

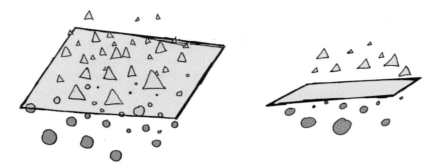

FIGURE 9.6   Cartoon of SVM classification in 3D with 2D hyperplane. It is difficult to visualize, but an SVM classifier can separate groups in much higher dimensions than three.

of the two-dimensional (2D) cartoon of Fig. 9.6, the hyperplane is the line dividing the groups. If the groups occupy three dimensions, then the hyperplane dividing the groups would be a 2D plane, much in the same way that a glass window can separate people on the inside from those outside. In dimensions higher than three, the visualization becomes a challenge.

In the following example, we use a *radial basis function* (RBF), which is a two-dimensional *Gaussian*, as our kernel to classify groups, the result of which contains remarkably radial outlines between the two groups. Note that the use of a nonlinear kernel in many cases is overkill: a linear separator between two groups will separate them cleanly, whereas a kernel which can take many shapes (depending on how you define the kernel) often requires more assumptions about the data. In particular the *kernel trick*, as it is called, is technically a method for assessing separation in a much higher dimension than that in which the separating hyperplane lives. The gist of the kernel trick is that you can separate groups in higher dimensions with functions more complicated than just straight lines, noting that this implicitly assumes much higher dimensionality of the data (Fig. 9.7).

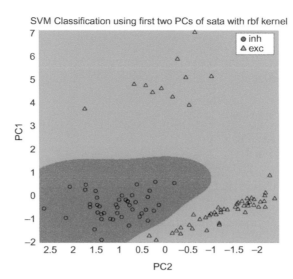

FIGURE 9.7 Voila: We built a support vector machine to classify the cells with a (Gaussian) radial basis function—it surely does look very circular. Your mileage might vary, as the SATA is created randomly in chapter 8. Below we test how well this division of groups generalizes to unseen data.

Pseudocode
```
 1. Import support vector machine package
 2. Create a support vector machine called clf
 3. Fit the model to the variables xTrain and yTrain
 4. Create figure
 5. Specify marker shapes: circle and triangle
 6. Specify marker colors red and cyan
 7. Specify marker labels for inhibitory and excitatory cells
 8. Specify grid resolution
 9. Create a colormap based on the markercolor variable
10. Determine the absolute maximum value of the xTrain array
11. Set the minimum plotting parameter to the negative of the max
12. Create a meshgrid that covers all x and y values. This meshgrid allows us to predict what group
    that any range of data would belong to
13. Predict what the support vector machine thinks the values for each x and y in the meshgrid should be
14. Reshape the predictions so we can plot them
15. Plot the meshgrid as a contour
16. For each value in the yTrain
17. Scatterplot the groups with the parameters specified by the marker values above
18. Set the x and y labels
19. Set the title
20. Show the legend
```

(Continued)

Python
```
from sklearn import svm #1
clf = svm.SVC() #2
clf.fit(xTrain, yTrain)#3
fig=plt.figure(figsize=(6,5)) #4
markershape = ['o','^'] # 5
markercolor = ['red','cyan'] #6
markerlabel = ['inh','exc'] #7
gridRes=.02 #8

cmap = ListedColormap(markercolor[:len(set(yTrain))]) #9
maxval = max(-1*xTrain.min(),xTrain.max())+.1 #10
minval = -maxval #11
xgrid,ygrid=np.meshgrid(np.arange(minval,maxval,gridRes),np.arange(minval,maxval,gridRes)) #12
Z = clf.predict(np.array([xgrid.flatten(),ygrid.flatten()]).T) #13
Z = Z.reshape(xgrid.shape) #14
plt.contourf(xgrid,ygrid,Z,alpha=0.4,cmap=cmap) #15

for yind,yval in enumerate(set(yTrain)): #16
    plt.scatter(x=xTrain[yTrain==yval,0], y=xTrain[yTrain==yval,1], c=markercolor[yind],\
    marker=markershape[yind], label=markerlabel[yind],s=50) #17
plt.xlabel('PC1');plt.ylabel('PC2')#18
plt.title('SVM First two principal components of sata with '+clf.kernel+' kernel') #19
plt.legend() #20
```

MATLAB
```
svmModel = fitcsvm(xTrain,yTrain,'KernelFunction', 'rbf'); %2 %3
classifiedFlat = predict(svmModel,[xFlat yFlat]); %13
classifiedGrid = reshape(classifiedFlat,size(xGrid,1),size(xGrid,2)); %14
figure %4
h = surf(xGrid,yGrid,classifiedGrid); %16
view([-90 90]);
h.LineStyle = 'none';
xlim([xMin xMax]); ylim([yMin yMax]) %11
hold on
h1 = plot3(rotatedSataArray(toPlot1,1),rotatedSataArray(toPlot1,2), ... %17
2*ones(length(toPlot1)),'o','markerfacecolor','r', 'markeredgecolor','k');
h2 = plot3(rotatedSataArray(toPlot2,1),rotatedSataArray(toPlot2,2), ... %17
2*ones(length(toPlot2)),'^','markerfacecolor','c','markeredgecolor','k');
legend([h1(1) h2(1)], {'inh','exc'}) %20
xlabel('PCA 1'); ylabel('PCA 2') %18
cMap = [1 0 0; 0 1 1]; colormap(cMap); %9
title('SVM Classification using first two PCs of sata with rbf kernel') %19
```

Other *kernel* types are available than the RBF. We could have used a *linear* kernel, in which one would expect an outcome very much similar to the logistic regression, which yields essentially a line (although the slope would likely be different). We also could have used a *polynomial kernel*, which is useful when separation of groups requires higher-order polynomials and would result in some kind of optimal squiggly line that optimally separates the classes of points.

Before moving on to *unsupervised* clustering methods, we want to discuss one final supervised classification method: random forests (Fig. 9.8).

They are somewhat beyond 1, so don't freak out if you don't get this immediately. The reason why we cover them here is that you should have an intuitive understanding of what they are and how to use them because they are becoming so important in data science.

Random forest regression is an *ensemble method* in machine learning (Breiman, 2001). Ensemble methods aggregate the results from many classifiers in an attempt to make the best possible predictions. Ensemble approaches have become quite popular.

Where do these forests come from? A forest is a collection of trees. But where do the trees come from?

In other words, a random forest lets you see the forest in spite of the trees—it gives you a big picture idea of what is going on.

FIGURE 9.8 Random forest.

Random forest regression specifically uses an ensemble of *decision trees* to form a model that is capable of making predictions about data (Fig. 9.9). The inner workings of decision trees require parsing the data into groups and testing how well these groups (or features of these groups) can account for variability in the data. The groups get divided into subgroups, and these subgroups get divided into subsubgroups, and so on, so that if we were to map out each extra division of groups, the result would look rather tree-like. At each branching of the data, we test how well the new subgroup can account for variance in the overall data.

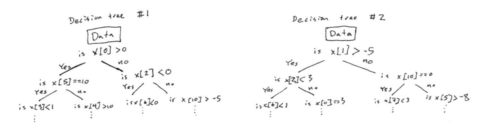

FIGURE 9.9   Example of two decision trees. The depth of the trees determines how many branchings occur. The random forest regression uses multiple decision trees, ultimately selecting parameters that generally fit the data.

How to divide up the data and test subgroups is an interesting question, and one that can result in *overfitting* of the data (wherein we would lose the power of predicting the correct class of previously unforeseen data). One way to prevent overfitting is to limit the number of branch points in the decision tree, known as the *depth* of the decision tree. Another way to prevent overfitting is to test many decision trees, which is exactly what the random forest regression does. The idea is that many decision trees are implemented and each of their predictive powers is tested. Each decision tree uses different branching points, so that if one of the trees happens to overfit the data, then we know it does not have good predictive power. We use an average of all the decision trees to figure out what the best general solution is to splitting up the data and predicting groups. The resulting regression of the data using many randomly split up decision trees is the aptly named random forest regression (Fig. 9.10).

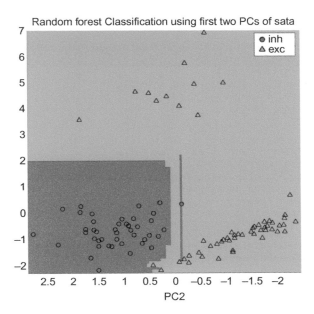

FIGURE 9.10 Random forest regression to classify types of neurons. This method can yield what looks like irregularly divided classes. Under the hood, the random forest method tests a bunch of twists and turns of subgroups of the data, before eventually resolving on what might appear to be a very nonlinear grouping to give us these classes. Which is what we see: unlike the logistic regression, which divides the groups by a straight line, the random forest can look like an Etch-a-Sketch divided the groups. This gives us the flexibility needed to classify more complex data. Here, this is overkill—the sata behaves nicely, as we knew it would (because we made it).

Pseudocode	1. import package for random forest regression
	2. create the random forest model
	3. fit the model with xTrain and yTrain
	Note that the code from here is the same as the above two example (except for the title)

(*Continued*)

| Python | ```
from sklearn.ensemble import RandomForestRegressor #1

regressor = RandomForestRegressor(n_estimators=10, min_samples_split=2) #2
regressor.fit(xTrain, yTrain) #3
fig=plt.figure(figsize=(6,5))
markershape = ['o','^'] # circle and triangle
markercolor = ['red','cyan']
markerlabel = ['inh','exc']
gridRes =.05

cmap = ListedColormap(markercolor[:len(set(yTrain))])
maxval = max(-1*xTrain.min(),xTrain.max())+.1
minval = -maxval
xgrid,ygrid=np.meshgrid(np.arange(minval,maxval,gridRes),np.arange(minval,maxval,gridRes))
Z = regressor.predict(np.array([xgrid.flatten(),ygrid.flatten()]).T)
Z = Z.reshape(xgrid.shape)
plt.contourf(xgrid,ygrid,Z,alpha=0.4,cmap=cmap)

for yind,yval in enumerate(set(yTrain)):
    plt.scatter(x=xTrain[yTrain==yval,0], y=xTrain[yTrain==yval,1], c=markercolor[yind], marker=
markershape[yind],label=markerlabel[yind],s=50)
plt.xlabel('PC1')
plt.ylabel('PC2')
plt.title('Random Forest First two principal components of sata with '+clf.kernel+' kernel')
plt.legend()
plt.savefig('Figure 9.10. Random Forest Regression of first two principal components of 100
neuron sata.png',dpi=300)
``` |
|---|---|
| MATLAB | ```
randomForest = TreeBagger(100,xTrain,yTrain); %2 %3
classifiedFlat = predict(randomForest,[xFlat yFlat]); %13
classifiedGrid = reshape(str2num(cell2mat(classifiedFlat)),...
                size(xGrid,1),size(xGrid,2)); %14
figure %4
h = surf(xGrid,yGrid,classifiedGrid); %16
view([-90 90]);
h.LineStyle = 'none';
xlim([xMin xMax]); ylim([yMin yMax]) %11
hold on
h1 = plot3(rotatedSataArray(toPlot1,1),rotatedSataArray(toPlot1,2), ... %17
2*ones(length(toPlot1)),'o','markerfacecolor','r', 'markeredgecolor','k');
h2 = plot3(rotatedSataArray(toPlot2,1),rotatedSataArray(toPlot2,2), ... %17
2*ones(length(toPlot2)),'^','markerfacecolor','c','markeredgecolor','k');
legend([h1(1) h2(1)], {'inh','exc'}) %20
xlabel('PCA 1'); ylabel('PCA 2') %18
cMap = [1 0 0; 0 1 1]; colormap(cMap); %9
title('Random Forest Classification using first two PCs of sata') %19
``` |

We can reliably determine whether the cells in our sataset are *excitatory* or *inhibitory* using these classification methods. In this example, we know that we can train our model to make a prediction of the type of cell any neuron in the sataset should be classified as, given the features we measured—latency, maximum spike rate, soma volume, etc. Because in the sataset we created, we know if the cells are excitatory or inhibitory. We thus can label them as 1 and 0, respectively. With these methods, we are able to create a model to predict whether neurons can be classified into known cell types. In general, such methods are called *supervised learning* methods. The name *supervised* implies here that we know the "answers," i.e., the 1's and 0's of the transmission cell type, which will guide the model to figure out what some new data should be classified as.

So let's do that. Let's use the models we built to make some predictions.

PREDICTIONS, VALIDATION, AND CROSSVALIDATION

What good is all this classification and clustering? And how do we know any of this is true? How do we know the machine didn't just overfit our data?

Here we introduce the concept of a test set. This entire time, we've been sampling 100% of the sata to build our model, using the variable trainPercent. This is called the "training set." We can also deliberately leave 10% of the sata out of this construction process. This left out data is the test set: where we put our classifiers to the test to see how well they can make predictions about data that were not involved in the creation of the model. This is useful as we can validate how well the classifiers group the data, as in the original sataset we have excitatory and inhibitory types. With the data left out, we can test how well the classification methods work on the remaining 10 cells. Recall from Chapter 7, Regression, that this is critical—if we used 100% of the data to build the model, we would be double-dipping and almost certainly overfitting if we used the same data to test the model, giving us a false sense of confidence (Fig. 9.11)

See companion site online for testing code: http://booksite.elsevier.com/9780128040430.

This is where *crossvalidation* enters the picture. The goal of crossvalidation is to reduce the possibility that we are *overfitting* the data. This is because we do not know if a single predictive model that we build will only work on the particular observations we made of the data we recorded. The ultimate scientific method for testing if a model overfits the data is to go out into the world, record more data, and see if the new data are consistent with the old data. As neural data scientists though, we have the foresight to simply set aside some of our data, and put these *held out* data to the test later on. The choice of which subset of data to hold out may itself bias the outcome, so we can iteratively hold out different subsets of data, each time forming a model on the remaining dataset, and each time forming a prediction. If each of these predictions results in tests that consistently verify the predictions of the model, then the models are not overfitting the data.

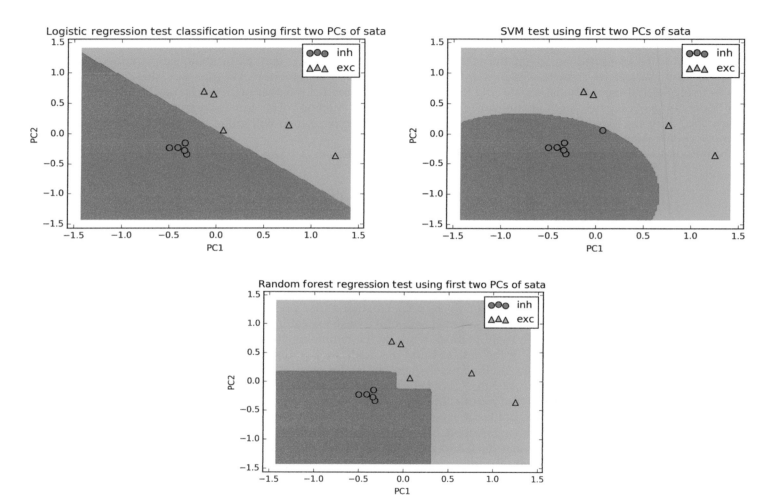

FIGURE 9.11   Testing the predictions of each classifier. Each classifier predicts which group each held out cell should belong to. For this analysis, we *train* the model using 90% of the sata, and then *test* how well the model does on the other 10% of the data.

A typical crossvalidation routine involves leaving out 10% of the data for testing, and forming a prediction on 90% of the data for training. Which 10% of the data you select though is rotated through the data. After rotating through 10 times, you end up with 10 predictions. If any subset of the data result in overfitting, crossvalidation will show poor predictive power in the subset. Using 10 iterations of testing isn't required though: more generally, you can do any number (say, k) of iterations. This is known as *k-folds crossvalidation* Fig. 9.12.

The code for crossvalidation of the above methods can be found on the companion website in POM.

1.

0-10	10-20	20-30	30-40	40-50	50-60	60-70	70-80	80-90	*90-100*

2.

0-10	10-20	20-30	30-40	40-50	50-60	60-70	70-80	*80-90*	90-100

3.

0-10	10-20	20-30	30-40	40-50	50-60	60-70	*70-80*	80-90	90-100

...

10.

0-10	10-20	20-30	30-40	40-50	50-60	60-70	70-80	80-90	90-100

FIGURE 9.12 Typical crossvalidation method involves shifting the 10% of data held out for testing, shifting 10 times, testing how well the other 90% of the training data predict the 10% held out. The 90% training data are noted as the **bold** numbers, the 10% testing data are noted as *italics*.

CLUSTERING

So far, we have only covered supervised learning. Supervised learning is kind of cheating. We already know the answers—in other words, which groups of neurons exist and how they are classified. The real business of a cognitive system is to figure out classifications of data in an unsupervised fashion. This is presumably the position that the brain is in for most of the time (outside of socializing forces like teaching, parenting, or culture)—think of the world as the blooming, buzzing confusion encountered by the brain of a baby (James, 1890). Unsupervised learning is about determining the underlying structure of the data from the data itself. This is called clustering. In cluster analysis, "correct" outcomes are not known ahead of time. We cannot use a training sample from a subset of the data to create a model in which we know how the cells should be grouped.

People are especially good at this kind of thing, particularly visually (Lintott et al., 2008).

There are many existing clustering methods and more are in development, as this is an area of active research. The one method we will introduce here is *k*-means clustering, as it gets us to "1" in clustering—you should know about it and understand what it is. It also nicely illustrates some of the basic shortcomings and challenges of clustering methods. As this chapter (and this book!) is already getting too long, the code for everything we discuss in the rest of the chapter can be downloaded from the companion website in POM. This is also efficient because the function that does all the work, e.g., `kmeans` in MATLAB is straightforward, but producing the figures below requires a lot of scaffolding code. Better to focus on the concepts.

The idea behind `kmeans` is quite simple. You specify the number of clusters you want to extract from data and the `kmeans` algorithm finds that number of clusters in the data. What the algorithm minimizes is the summed distance between each data point and its closest cluster center. Cluster centers are moved around until the overall distance from all data points to all cluster centers is minimal. As distance minimization drives the `kmeans` algorithm, there are several distance metrics one can use, most straightforwardly euclidian distance (as the crow flies), cityblock (think of walking around Manhattan—aerial distance will be of limited use, unless you are a drone), and others. In other words, the `kmeans` clustering algorithm tells you *where* cluster centers (centroids) are, if there is a certain number of them, but not how many of them there are. Ideally, this is something you would get the algorithm to tell you, figured out only from the data itself, not something you have to tell the algorithm.

An important implication of this is that `kmeans` will always find as many centroids as you ask it to. That doesn't mean they are "real." The problem is that in any actual dataset, you probably won't know the number of "real" clusters beforehand. Usually, this is something you want the data to tell you, via an algorithm.

To show you that this is true, let's do `kmeans` on our sataset. We know that there are six distinct types of neurons because that's how we created the sata. But if we ask `kmeans` to find two clusters in the data, it will do so and if we ask it to find 10, it will also do so (Fig. 9.13):

As you can see, `kmeans` is quite obliging. If you ask it to put any number of centroids on your data, it will do so. Put differently, `kmeans` is great at putting the position of centroids if you know how many there are. If you don't know how many there are, `kmeans` might get you in trouble. As a matter of fact, you could ask `kmeans` to find as many cluster centers as there are datapoints in the sample, and it will do so, perfectly matching each datapoint, but also perfectly overfitting this particular dataset.

One way to overcome this problem is to do a "silhouette analysis" (Rousseeuw, 1987). The point of silhouette analysis is to determine how many clusters are likely to be valid. Briefly, `kmeans` itself only takes information about distances of datapoints from the nearest cluster into account. As we have seen, a datapoint is deemed to belong to the cluster with the closest centroid. In silhouette analysis, we also consider the distance to other cluster centers. Specifically, we calculate the average distance of a datapoint and all other data in the same cluster as well as the average distances of the datapoint to the data in all other clusters. The neighbouring cluster is defined as that cluster with the lowest average distance of the

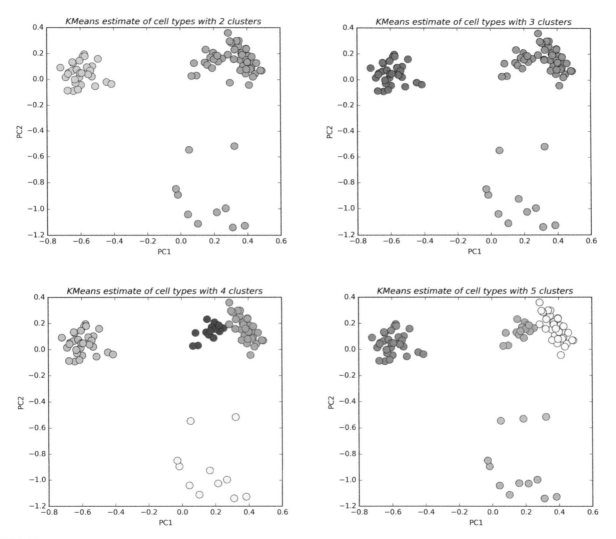

FIGURE 9.13 Kmeans clustering of our sataset with number of clusters between two and ten.

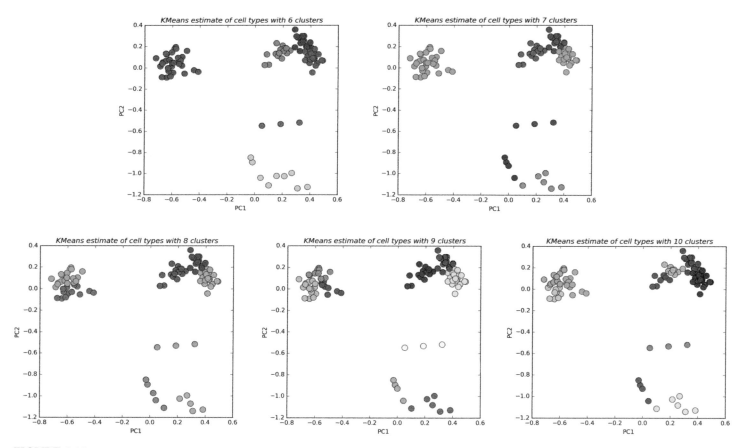

FIGURE 9.13 (*Continued*)

other clusters. The "silhouette" is then defined as the average distance to the data in the neighboring cluster minus the average distance to the data in the same cluster, standardized by the largest average distance of same and neighboring clusters. This yields a silhouette coefficient between −1 and 1. A value of close to 1 indicates that the point is quite representative (or close to) the other data in its cluster and quite far away from data in the neighboring cluster. A value around 0 indicates that the classification is almost arbitrary, as the point is close to the decision boundary between two clusters. A negative value indicates misclassification. So you could run the kmeans clustering multiple times, do the silhouette analysis in each case and pick the number of clusters as valid that has the highest average silhouette score (Fig. 9.14).

As you can see, silhouette analysis suggests that two clusters is the "real" number of clusters in the first two principal components of this sataset, even though the number of neuron types in the sataset was 6. As this is sata, created by us, we know that this is true. To make the classification work, we decided that only whether the neuron is excitatory or inhibitory should matter. But the most suitable number of clusters as suggested by silhouette analysis must not be the smallest. To illustrate this point, we did an analysis on different sata where the true number of clusters is 4 (see figure 9.14b). You can download the code that produces this figure from the companion website ("kmeansDemo") But what can we do with this? For instance, suppose we did know that there are six cell types in our sample, it might be misleading for the purpose of figuring out the nature of the local microcircuit to lump them all together. It would be better to calculate sample means (for instance, in response to a stimulus) per cell type, not overall. But we only have a handful of neurons in our sample for some of the cell types. The central limit theorem suggests that we need at at least 20 instances of a group (more would be even better) to achieve a statistically reliable estimate of the sample mean of neurons in that group. But we don't have 20 for some of these. What is a data scientist to do?

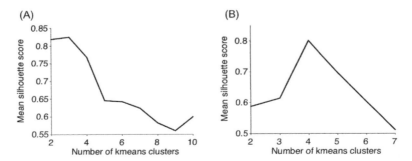

FIGURE 9.14   Mean silhouette scores as a function of number of clusters. (A) From the first two principal components of our sataset. (B) From a demo with 4 known and clearly separated clusters.

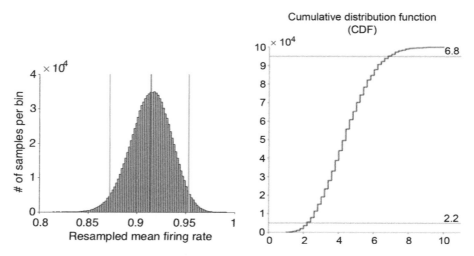

FIGURE 9.15   The 95% confidence interval of firing rate sample mean for cell type 6, based on a million resampled samples. Red: Resampled mean. Green: Upper and lower CI bound.

In a nutshell, bootstrap (Efron & Tibshirani, 1994).[96] The idea behind the bootstrap is that we sample (with replacement) many times (like a million times) from the data we do have to construct a confidence interval of the sample mean, which is all we need to assess the stability of the sample mean. This might seem miraculous, but it works reasonably well in many cases (Fig. 9.15) Note that it is controversial to bootstrap in cases of small sample sizes, as it—by definition—will under- or overestimate the prevalence of rare cases if the small sample is not representative of the population, which a small sample is unlikely to be. Bootstrapping works best if the sample is fully representative of the population because we effectively use the sample as a stand-in for the population. If the sample is biased or unrepresentative, the result will also be biased. In any case, the code that produced figure 9.15 is online, on the companion website.

What if we apply the techniques from Chapter 8, Dimensionality Reduction and this chapter to real data?

Using the data from Chapter 4, Correlating Spike Trains, from the Smith lab, we, as a first pass, use PCA on the normalized PSTHs and visualize a scatter plot of the first two principal components for each unit (Fig. 9.16).

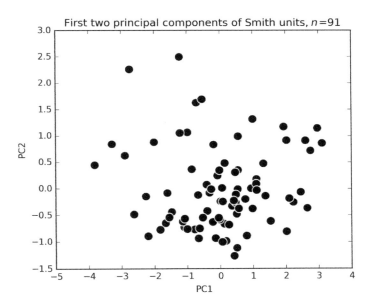

FIGURE 9.16 PCA on PSTHs from the Smith lab data.

Our immediate impressions of using PCA on the real dataset are that we do not observe clearly delineable groups based on these two components alone. This is not surprising: nearly a hundred cells, drawing from many different firing types deep within the neural cascade—of course it wouldn't be this easy. To illustrate just how complex the problem is, we look at just how well the principal components describe the variance in the data. As we predicted in Chapter 8, Dimensionality Reduction, the scree plot in Fig. 9.17 shows that to account for 90% of the variance, we must use nearly 30 principal components.

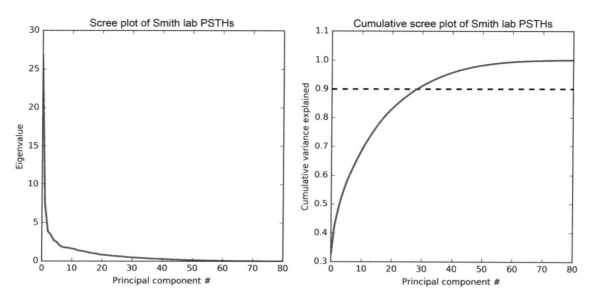

FIGURE 9.17   Scree plot of Smith lab dataset.

⌘. *Pensee on data analysis:* If you have recorded any data, you will know that data are hard won. The age-old question whether a tree makes a sound if it falls in a random forest, but no one is there to observe it has been resolved. The falling tree will create a pressure wave in the air as it displaces air molecules, but unless these impinge on the eardrum of an auditory system, and these signals are interpreted by a nervous system, there won't be any sound. As a matter of fact, which sound is heard depends on the auditory system of the organism present—an elephant might hear it strongly due to the low frequencies in the signal whereas a bat might not perceive it at all (Von Uexküll, 1992; Nagel, 1974; Heffner & Heffner, 1985). The hearing range of mice and men is somewhere between those extremes. Data take this concept a step further. Even if a natural process is observed, it is not data unless it is also recorded. That's a lot to ask for. That's why one should make the most out of data. In other words, do someone's data analysis for them, and they will be ok for a day, but teach them how to bootstrap, and they will never go hungry. That's what we hope to accomplish here.

Web Scraping

In prior chapters, we dealt exclusively with data and sata, mostly from physiological sources. In this final chapter, we go online and enter a world full of *metadata*. Many data science applications today require the using or harnessing of real-time streams of social or financial data. If an engineer wishes to obtain data from a website, they use one of two general approaches: they can either build their own custom web-scraper (Python has many libraries for this, including `urllib2` and `beautifulsoup`), or, if the engineer is lucky, they can use an application program interface (API), which, in the case of retrieving data from the web, would be in the form of a package written specifically for retrieving data from that website. APIs can take many shapes and forms: e.g., if you are interested in online sales, Amazon has a Product Marketplace API, if you wish to use Google Maps for your own map creation tools—you got it, there's an API for that too.

For this chapter, we will be querying data found on the National Institutes of Health's repository for publication listings, *PubMed*. We could use `urllib2` or many other packages to build a custom webscraper to play with abstracts and publication data, but fortunately, the good people over at biopython have provided us (and the world) with Entrez, an API for querying data on PubMed (see `http://biopython.org/DIST/docs/tutorial/Tutorial.html`).

Note that MATLAB can also scrape websites, using the function `urlread`. This function returns the actual html of the requested website as one long string. In principle, we could parse this string vector for what we are looking for, but this is a book about going from 0 to 1—doing web scraping is well beyond 1 in MATLAB. Some things are beyond 1 in Python, e.g., some sound handling and some signal processing, as we saw in Chapter 5, Analog Signals. As one should generally use the most suitable tool for the job, we suggest to just sticking with Python on this one—that's what we did. Web scraping is arcane enough as it is—it is an important data science use case (which is why we cover it in this final chapter), but a sharp deviation from anything we've done so far.

Specifically, in this chapter we will try to ask the question: *Has the use of colons (:) increased in the titles of publications over time?* Anecdotally, there is reason to believe that this is the case in some fields. As the number of publications is ever increasing, titles assume more and more the role of advertisements - it is the only thing that most people will ever read of any given publication. As such, purely technical titles are no longer sufficient and many authors now adopt the general format of: *Eye catching title: Technical title*. Thus, we would predict that the ratio of publications with colons in the title has increased over time. Can we substantiate this hunch with data?

Pseudocode	1. Import packages To get Entrez you may need to run the command (from the command terminal, not within Python): pip install biopython Or conda install biopython

Python	```from Bio import Entrez #1```
	```from collections import defaultdict```
	```import matplotlib.pyplot as plt```
	```from scipy import stats```
	```import numpy as np```

Pseudocode	1. Define search term 2. Define Character of Interest to be ":" 3. Return maximum number of publications

Python	```searchterm = 'psychopathy' #1```
	```COI = ':' #2```
	```maxReturn=10000 #3```

The Entrez API first requires you to give your email. We then create a handle in which we query the "pubmed" database (there are other options for what to query), as well as the search term, which in this case is "psychopathy" (this could be anything, like money priming, power posing or cortical motifs) and the maximum number of results to return, which we specified with the variable `maxReturn` to be 10,000. The method to perform this query is specific to Entrez, called `esearch`. All APIs have their own rules as to how queries are handled. If you are lucky, they will be well documented. With the handle, we can then read out the results, which we hold in the dictionary record. Note that we used "psychopathy" here as an example because of the research interests of one of the authors. If you study something else and want to know what the trend in your field it, input another search term instead.

Pseudocode	1. Always tell NCBI who you are 2. Search PubMed according to searchterm 3. Retrieve record dictionary 4. Extract idList

Python	```Entrez.email = "neuraldatascience@gmail.com" #1```
	```handle = Entrez.esearch(db="pubmed", term=searchterm,retmax=maxReturn) #2```
	```record = Entrez.read(handle) #3```
	```idlist = record["IdList"] #4```

In this dictionary, `record` contains the values specified by the key `"IdList,"` which is, as it sounds, a list of PubMed IDs for all the publications that contain the search term "psychopathy." We use this list of IDs to make another query with Entrez using the method `efetch`. We specify the pubmed database, input the list of IDs, specify medline, and then also specify the datatype returned to be XML (a notoriously cumbersome data type, but, we're used to handling neural data, so how bad can it be?).

We then read the `records` from this into the variable `records`.

Pseudocode	1. Fetch data based on the idList 2. Read out the records. Note you may receive the error HTTP Error 403: Forbidden for trying to retrieve too many files.
| Python | ```
handle = Entrez.efetch(db="pubmed", id=idlist,
rettype="medline", retmode="xml") #1
records = Entrez.read(handle) #2
``` |

We now employ some highly customized Python code to break into these records. One common problem in data munging, especially with new data, is figuring out just what kind of object or type you are dealing with. A few things you can do to view types are:

```
>>> type(records)
Bio.Entrez.Parser.ListElement
```

or use the function `dir()` which returns a list of the directory of whatever it is you are interested in. If after this book you find yourself diving deeper into the Python programming language, you'll definitely learn about how everything in Python is an object, this directory is your map to every object's structure, and how all objects within objects are inherited. Try using `dir()` for yourself:

```
>>> dir(records)
['__add__',
 '__class__',
 ...
 '__weakref__',
 'append',
 'count',
 'extend',
 'index',
 'insert',
 'pop',
 'remove',
 'reverse',
 'sort',
 'tag']
```
```

The stuff at the end here is all things that mean records is indeed a list (as suggested by the type() above). The natural thing then is to look at the first element of the list:

```
>>> records

{u'MedlineCitation': DictElement({u'OtherID': [], u'OtherAbstract': [], u'CitationSubset': ['IM'],
u'KeywordList': [], u'DateCreated': {u'Month': '06', u'Day': '24', u'Year': '2016'}, u'SpaceFlightMisson':
[], u'GeneralNote': [], u'Article': DictElement({u'ArticleDate': [], u'Pagination': {u'MedlinePgn': '772-
80'}, u'AuthorList': ListElement([DictElement({u'LastName': 'Saha', u'Initials': 'TD', u'Identifier': [],
u'AffiliationInfo': [{u'Affiliation': 'Laboratory of Epidemiology and Biometry, Division of Intramural
Clinical and Biological Research, National Institute on Alcohol Abuse and Alcoholism, National Institutes of
Health, 5635 Fishers Ln, Room 3083, Rockville, MD 20852. sahatd@mail.nih.gov.', u'Identifier': []},
...
, attributes={u'PubStatus': u'medline'})]}}
```

This is what we meant by XML being a mess. But that's OK, we figured out how to get the good stuff out of here, and the reality is that the record itself is just a bunch of nested dictionaries, with the occasional odd formatting wrench thrown in there. Each key in the dictionary here is usually a string, though the strings above have a "u" in front of them. Worry not—this just means the string itself is *unicode*, which means it follows a certain standardized system of representing symbols from human languages (if you haven't encountered unicode before and are looking for a good Internet hole to go down, Google *unicode*—you are likely to encounter many scripts from cultures that you were entirely unaware of).

Now we try cycling through each record to see if the *character of interest,* noted as variable COI, the colon ":" is contained in the title. We do this by retrieving the title string, and then splitting it by the colon. If we split a string by a character that doesn't exist, a list with the single original string is returned. However, if the string contains a colon, and we split the string by the colon, returned will be a list of the strings around which used to wrap the colon (and the colon will be gone). Thus, the length of this output list can be used to determine whether or not the colon is in the title. If we did this in MATLAB, we would simply read in each title and either go through it character by character, each comparing to a colon with strcmp (and stopping once a colon is encountered) or look for the colon with the function regexp. The regular expression function returns the position in a larger string in which a piece of the string is found. This would work ideally in this case.

Back to Python. For each record, we add whether (1) or not (0) a colon is in the title to the dictionary yearcolondict.

We are also aware that records are not perfect. Sometimes, there will be inconsistencies in the way data are stored, particularly when handling hundreds of thousands of entries from various sources. So it is prudent to anticipate this and create code that can handle errors and doesn't just stop when encountering one.

The function to do this in POM is try, which is paired with except in Python, and with catch in MATLAB. The function try will attempt to execute the code within. However, if an error is encountered, instead of the program throwing an error and stopping, it will execute what is contained in except/catch. Again, we add this condition here because of some inconsistencies in how the year is stored in the data. We need to know the publication year in order to answer our question of whether the proportion of colons in titles has increased or not. The API restricts the number of records we can pull at once. But this is no problem. Let's return a small number (100) for each query, and pull the data year by year, instead of all of them at once.

Pseudocode	1. specify the dictionary outside of looping through the years
	2. for each year between 1970 and 2015
	3. specify the max number returned as 100, in case Entrez has issues with you pulling thousands at a time
	4. give them your email
	5. search with the parameters where minimum and maximum date are the year specified
	6. read out the records
	7. get the IDs of that record
	8. fetch the full records according to the IDs
	9. read out the records
	10. for each index and record in records
	11. try the next line (in case there are errors in the XML)
	12. if the splitting the title on a colon gives a list of length greater than 1 (if there is a colon in the title)
	13. append a 1 to the dictionary for the year of interest
	14. if there is no colon in the title
	15. append a 0 to the dictionary for the year of interest
	16. if the previous statement failed try these other commands
	17. print out a statement specifying that there is no date info
	18. try out the following code
	19. if that doesn't work
	20. print an error statement

Python

```
yearcolondict = defaultdict(list) #1
for year in np.linspace(1970,2015,46): #2
    maxReturn=100 #3
    Entrez.email = "neuraldatascience@gmail.com" #4
    handle = Entrez.esearch(db="pubmed",
term=searchterm,retmax=maxReturn,mindate=year,maxdate=year) #5
    record = Entrez.read(handle) #6
    idlist = record["IdList"] #7
    handle = Entrez.efetch(db="pubmed", id=idlist,rettype="medline", retmode="xml") #8
    records = Entrez.read(handle) #9
    for ind,record in enumerate(records.values()[0]): #10
        try: #11
            if len(record["MedlineCitation"]["Article"]["ArticleTitle"].split(COI)) > 1: #12
                yearcolondict[year].append(1) #13
            else: #14
                yearcolondict[year].append(0) #15
        except: #16
            print 'no date info for ind: ',ind #17
        try: #18
            if len(record["MedlineCitation"]["Article"]["ArticleTitle"].split(COI)) > 1: #12
                yearcolondict[year].append(1) #13
            else: #14
                yearcolondict[year].append(0) #15
        except: #19
            print 'the next try statement didnt work...' #20
```

Now we have a dictionary `yearcolondict`, which contains keys for each year, and values which are lists of 1's and 0's, indicating whether or not the title contained a colon. We thus cycle through each year of records, calculate the total number of colons that year, divide by the number of publications that year, and plot as a function of year. We append the colon to publication ratio to the variable x and the year to variable y.

We use x and y to, while we're at it, do a linear regression to these data and fit the resulting trend with a line so we can quantitatively describe the trend in publication naming. We'll use Python's `scipy.stats.linregress`, which returns the `slope` and `intercept` of the trend (in addition to the r-value, p-value, and standard error) (Fig. 10.1).

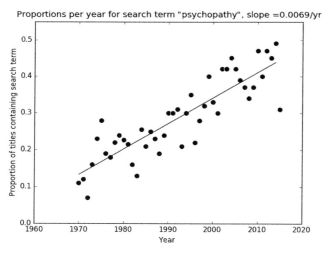

FIGURE 10.1   The ratio of colons in publication titles each year for articles that contain the search term *psychopathy*.

Pseudocode	1. Create figure
	2. Create subplot
	3. Initialize x and y variables
	4. For each key in yearcolondict (will be a list of years)
	5. Calculate the total number of colons that year
	6. Calculate the total number of publications that we measured from that year
	7. Scatter plot of proportion of colons vs year of pubs
	8. Append the year to x
	9. Append the proportion to y
	10. Calculate a linear regression between x and y
	11. Create a new range of evenly spaced x-values
	12. Create y-values based on the linear regression (and the new range of x-values)
	13. Plot the new y-values vs the x-values
	14. Set the title
	15. Set the xlabel
	16. Set the ylabel
	17. Set the ylimits
	18. Save the figure, since you'll want to remember this

Python

```python
fig = plt.figure() #1
ax = fig.add_subplot(111) #2
x=[];y=[] #3
for year in yearcolondict.keys(): #4
    totescolons = float(sum(yearcolondict[year])) #5
    pubsthisyear = float(len(yearcolondict[year])) #6
    ax.scatter(year,totescolons/pubsthisyear,c='k',s=80,edgecolor='w') #7
    x.append(year) #8
    y.append(totescolons/pubsthisyear) #9
slope, intercept, r_value, p_value, std_err = stats.linregress(x,y) #10
xvals = np.arange(min(x),max(x),1) #11
yvals = slope*xvals + intercept #12
ax.plot(xvals,yvals,c='k') #13
ax.set_title('Proportions per year for search term "'+searchterm+'", slope
='+str(slope)[:6]+'/yr',style='italic') #14
ax.set_xlabel('Year',style='italic') #15
ax.set_ylabel('Proportion of titles containing search term',style='italic') #16
ax.set_ylim([0,0.55]) #17
plt.savefig('Figure 10.1 Character proportions per year for term '+searchterm+' char
'+COI+'.png') #18
```

It looks like we were right—the proportion of titles using the search term "psychopathy" has been steadily increasing since the 1960s. If you don't study psychopathy and are curious what the trend is in your field, try it. As we already saw in Chapter 7, Regression, linear regressions are of course great for making predictions. If we think the trend in use of colons in publication titles is linear (which we don't, but stay with us for a moment), we can predict the proportion of titles containing colons in the future. We have already explored in Chapter 7, Regression, that extrapolating over time, projecting into the future is a dangerous game to play, be it by assuming linearity or with exponential functions. The only case where this really worked well (to our knowledge) was Moore's law, and even that is starting to come to an end (Moore, 1965; Kumar, 2012; Waldrop, 2016). To further show that linear extrapolation can, and usually will, yield absurd results, we use this example to project the proportion of titles that will contain colons.

Pseudocode	1. project our regression to a year in the future 2. create figure 3. create subplot 4. for the year and index in x 5. scatter plot the year vs the y 6. create new x-values (including the future year) 7. create new y-values 8. plot the two 9. create a blue point at the future year 10. set the title 11. set the xlabel 12. set the ylabel 13. draw a dashed black horizontal line at y=1 14. save this figure

Python

```
year_project = 2030 #1

fig = plt.figure() #2
ax = fig.add_subplot(111) #3
for yearind,year in enumerate(x): #4
    ax.scatter(year,y[yearind],c='k',s=80,edgecolor='w') #5

xvals = np.arange(min(x),year_project,1) #6
yvals = slope*xvals + intercept #7
ax.plot(xvals,yvals,c='k') #8

ax.scatter(year_project, slope*year_project +
intercept,c='b',s=80,edgecolor='w') #9

ax.set_title('An obscene projection of use of colons in psychopathy research
to year '+str(year_project),style='italic') #10
ax.set_xlabel('Year',style='italic') #11
ax.set_ylabel('Proportion of titles containing search term',style='italic') #12
ax.axhline(1,ls='--',c='k') #13
plt.savefig('Figure 10.2 Projected character percents per year for term
'+searchterm+' char '+COI+' to year '+str(year_project)+'.png') #14
```

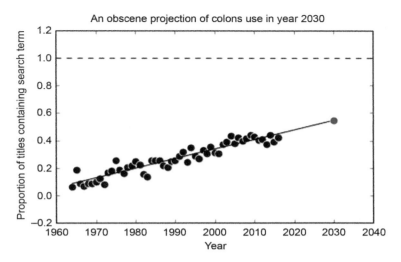

FIGURE 10.2   Projection with linear regression that may seem appropriate. Black dots: Data. Blue dot: Prediction.

Will more than half of papers on psychopathy in 2030 contain colons in the title (Fig. 10.2)? It's possible. How about in 2150 (Fig. 10.3)?

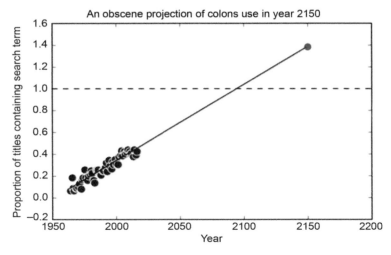

FIGURE 10.3   Projection with linear regression rendered absurd. Black dots: Data. Blue dot: Prediction.

Unless Kurzweil is right, we probably won't be here to tell whether the prediction came true or not. But we don't have to be. By definition, a proportion can't be larger than 1. In other words, the linear trend can't go on forever—and perhaps has stalled already. (if you look at the data past the year 2000, it has basically remained flat since then).

This simple exercise got us to 1 in web scraping, which any data scientist should be able to do. There is much more beyond 1. Note that if you absolutely insist to do all of this in MATLAB, you can of course do so. We'll leave this as an exercise for the reader, but the most straightforward thing to do is to use the Bioinformatics toolbox, which has a getpubmed demo (and functions to access various web databases). Also, try to use the new string manipulation functions that were introduced in 2016b such as "split", "startsWith", or "contains", which will obviate the need to use regexp as well as the more versatile "webread" function, which replaces "urlread".

WHAT LIES BEYOND 1?

Lots of things—this is just the end of the beginning, not the beginning of the end. Here is a rough idea of what you can expect.

Going from 1 to 10: You'll start learning more about objects in Python (namespaces and inheritance). You'll use Sublime Text editor and a command window, but also bounce between Jupyter Notebooks. You'll launch a web app using Flask and log your own behavioral data online. You'll version control your code and share with collaborators on GitHub. You'll read other books on programming in Python. You'll seek help from other neural data scientists.

From 10 to 100: You'll start using the Thunder package which utilizes Spark for parallel processing of large imaging data. You'll read papers on machine learning. You'll use Beaker Notebooks because you'll be doing some analyses in R and some visualization with D3 (you'll learn some JavaScript). You'll do a little data science consulting on the side. You'll know how to use Gradient Boosting. You'll compete on CodeNeuro. You'll contribute to open source projects. You'll present your findings to audiences at NIPS. You'll get neural networks, as much as it is possible to get neural networks. You'll run your code on Hadoop. You'll lift up those who are at zero and get them to 1 so that they may one day join you at 100.

From 100 to 1000: You will evangelize machine learning to your mortal friends. You will write everything in C if possible. You will build packages for Julia. You will teach incredibly dense concepts to those who otherwise would not be able to break into the box.[36]

Beyond 1000: We don't know. Ask Eero (Wickelgren, 2006) (Fig. 10.4).

FIGURE 10.4   The authors in Rutherford, NJ. Taken June 2016.

⌘. *Pensee on prospects and pitfalls:* We hope this book was a helpful guide on your beginning journey and will aid in reaching your next destination, whatever that might be. Importantly, we hope that we have managed to impart a philosophy of empowerment that was implicitly ever present throughout this text, but which we now want to make explicit, as we gather these parting thoughts. The point is that data represent—by virtue of how they came to be (through a systematic measurement process)—a relevant aspect of the natural world quantitatively. So you will encounter lots of numbers and lots of ways to do operations on them, yielding new numbers. In this sense, a data scientist can conceive of themselves as a number surgeon. But that's what everything in this book (and everything in data science) boils down to: Operations on numbers, some straightforward, others more complex. But all of them involve numbers and produce new numbers. That's it. Whatever these processes might be called—some of them can have intimidating names like the random forests or support vector machines we encountered in the last chapter. Hopefully, we managed to demystify these concepts a little bit—always think of what happens to the numbers, as the operation is performed on them, however fancy it might sound.

In the same vein, as most of the people reading this probably started with MATLAB, we hope that we defanged the snake a little bit. We recognize that Python can be a bit intimidating at first. But it is powerful, just like the actual animal.

Finally, we use this farewell address to warn of the pitfalls of tribalism. In 1618, Europe went to war. Over religion. Or rather, the interpretation of the same religion. At stake was essentially whether people need a mediator in interpreting the word of God or not. Thirty years later, large swathes of Europe (particularly in Germany) were ravaged, millions of people were dead—scaled by today's population, it would be a conflict with well over 60 million fatalities (Theibault, 1997), with population losses approaching 50% in locations at the center of the fighting. It was a devastating conflict. What settled the matter, leading to the peace of Westphalia that established the modern, nationalistic world order in 1648, was not arguments, but naked force. Today, as western societies become ever more secular, religion is less and less of a battlefield. But the tribalist circuitry in the brain that drove conflicts like the 30 Years War over matters that seem obscure to us today is still in place, ready to drive people into separate camps over things that should not really matter, such as which phone to use, which operating system and of course, also, which programming language.

But that misses the point. In contrast to religion, you don't have to pick. Knowing how to use more tools makes it more likely that you have the right one for the job. To someone who only knows MATLAB, everything looks like a matrix. To someone who only knows Python, everything looks like a tupel. This might not be optimal for any given use case. So try to overcome platform chauvinism, it keeps you back and is needlessly divisive. That's why we tried to teach two languages in parallel here. Don't feel the need to stick to that. Pick up some C, pick up some Java, pick up some Julia while you're at it. When it comes to programming languages, more is almost always better. You do not have to choose, you do not have to look down on others who made different choices. There is absolutely no need for that. Use tools, don't be one.

The point of this book was to help foster computational empowerment for neuroscience practitioners, allowing them to be able to do their own basic analysis and for data scientists to never lose touch with the measurement process that marned the data. That's it. Don't get bogged down in platform tribalism. It will just keep you back.

A

MATLAB to Python
(Table of Equivalences)

Whereas much of the syntax between POM is very similar, here are some of the most basic differences between MATLAB and Python.

COMMENTS

Good programmers use comments to make notes to themselves and to others who read their code. Comments are ignored by the machine when the script is executed. Python uses a hashtag to note comments, while MATLAB uses the percent sign. Note that in Python, you also have the option of enclosing comments with three scare quotes on both ends of the comment.

Python	MATLAB
# comments in Python use the hashtag	% comments in MATLAB use the percent
'''comments can also be made by enclosing the entirety of the statement in three scare quotes on both ends'''	

BLANKSPACE

Blankspace in Python is a requirement for any function definition, loop, or conditional. The return to a new line in Python indicates the end of that definition, loop, or conditional. In MATLAB, blankspace is not required for these, but is good practice for making code more readable and pretty ("Smart indent").

Python	MATLAB
``def print_odd(a):
for x in range(a):
if x%2:
print x```	```function print_odd(a)
for x = 1:a
if mod(a,2)
a
end
end
end```

LOOPS

In addition to the blankspace, note the subtle difference in syntax in for loops between Python and MATLAB.

Python	MATLAB
``>>>for a in range (3):
>>> print a
0
1
2```	```>>for a = 0:2
>>a
>>end
a=
0
a=
1
a=
2```

EXPONENTS

Python	MATLAB
>>>print 5**2	>>5^2
25	25

LISTS AND CELLS

In Python, dissimilar data types can be stored together in a list, created using square brackets. In MATLAB, dissimilar data types can be stored together in a cell, created using curly braces.

Python	MATLAB
>>>A= [5,'a',True]	>>A={3,'a',true}
>>>print A	A=
[5,'a',True]	[3] 'a' [1]

INDEXING

Python indices begin at 0. So the first element of list a is retrieved by calling a[0]. MATLAB indices begin at 1, so the first element of vector a is retrieved by calling a(1).

Python	MATLAB
>>>a=[4,9,10]	>>a=[4,9,10];
>>>print a[0]	>>a(1)
4	ans =
	4

Like with everything else in programming (e.g., tabs vs spaces), people feel strongly about the proper way to index—starting with 0 or 1. This divide has historical and philosophical roots. Most general purpose programming languages like Python or C index from 0, largely due to a more efficient handling of pointer arithmetic. In contrast, most high-level science and math-oriented languages like MATLAB, Mathematica, or Julia, index from 1. Indexing from 1 lends itself more naturally to many mathematical operations—e.g., one indexes from 1 to n, not from 0 to $n-1$. Also, the first element of an array can arguably be called element 1, not 0, and it is more natural to think of an array with 3 elements as having a length of 3, not 2. Then again, general purpose programming languages that index from 0 usually indicate the offset one needs to input when accessing a location in memory. This also has its advantages. For instance, the origin of a spatial map should probably be location 0,0, not 1,1. Similarly, 0-indexing is more natural for time-series and frequency space analysis. The first time we consider is usually time 0 and the first frequency is frequency 0.

Unfortunately, this is the source of a great many "off by 1" errors, but it seems like the divide is here to stay as most people just keep doing what they are used to and—as discussed earlier—there are legitimate reasons to use either convention. We advise being extremely mindful of this distinction to avoid commonly made errors.

IMPORTING PACKAGES VERSUS DEFAULT PACKAGES

MATLAB has most packages built right in: no importing of extra tools is needed. In Python, you must explicitly state which packages you will be using. You can give the packages nicknames too in order to save time later when typing the names. The packages `numpy`, `scipy`, and `matplotlib` are all very similar in functionality to what MATLAB has to offer.

Python	MATLAB
`>>>a=[1,2,3]`	`>>a=[1,2,3];`
`>>>import numpy as np`	`>>mean(a)`
`>>>print np.mean(a)`	`ans =`
`2`	`2`

RANDOM NUMBER GENERATION

Python spits out a row vector, MATLAB spits out a column vector.

Python	MATLAB
```>>> import numpy```	```>> rand(5,1)```
```>>> numpy.random.rand(5)```	``` 0.757740130578333```
```array([0.31075661, 0.45360828,```	``` 0.743132468124916```
```0.95271364, 0.14573102, 0.63847612])```	``` 0.392227019534168```
	``` 0.655477890177557```
	``` 0.171186687811562```

NUMERICAL TYPES

In Python, the *float* type is a generally useful numerical type, but if you need precision, the *decimal* function allows you to specify how many decimal points you want to store using *getcontext().prec*. Some of the Python syntax may appear foreign, for example *from decimal import* * is importing all functions from the function *decimal*, but note we will explain these in detail in Chapter 2, From 0 to 0.01.

Python	MATLAB
```>>> from decimal import *```	```>> format long```
```>>> getcontext().prec = 15```	```>> 1/7```
```>>> Decimal(1) / Decimal(7)```	``` 0.142857142857143```
```0.14285714285714285```	```>> single(1/7)```
```>>> getcontext().prec = 4```	``` 0.1428571```
```>>> Decimal(1) / Decimal(7)```	```>> format short```
```0.1429```	```>> 1/7```
```>>> getcontext().prec = 2```	``` 0.1429```
```>>> Decimal(1) / Decimal(7)```	```>> single(1/7)```
```0.14```	``` 0.1429 >>```
	```>> format bank```
	```>> 1/7```
	``` 0.14```
	```>> single(1/7)```
	``` 0.14```

Frequently Made Mistakes

Ninety-five percent of coding mistakes by beginners fall into of one of these 12 categories. Generally speaking, try to welcome the red ink, as frustrating as it might be at first: the software is telling you that it didn't understand one of your commands. The worst kind of error is the one where it is not doing what you think it is doing, but you are never alerted to this fact because the computer thinks it understands what you want. This can be quite dangerous. That said, error messages can be cryptic, so we are decoding the ones you are most likely to encounter.

1. *Not closing a parenthesis:* Every parenthesis that is opened must be closed again. Importantly, the kinds of parenthesis must match—Python and MATLAB distinguish three kinds: regular parentheses ( ), square brackets [ ], and curly braces { }. They all have different purposes.

   Example of a typical error and error message that you might see in MATLAB (note that you will get an error message that makes a guess as what you might have meant):

   ```
   >> A(3
   Error: Expression or statement is incorrect--possibly unbalanced (, {, or [.
   ```

   Or in Python:

   ```
   SyntaxError: invalid syntax
   ```

But don't add more than you need (they need to match exactly—not too many, not too few):

```
size(A))
Error: Unbalanced or unexpected parenthesis or bracket.
```

2. *Trying to create an illegal variable name:* Python and MATLAB is quite permissive when it comes to acceptable variable names, but there are three exceptions: they can't start with a number, there can be no blank space in the middle (it has to be one word), and they can't contain special characters.

   The kind of error message you will get depends on which of these rules you violate. In turn:

   a. Starting a variable name with a number, MATLAB politely tells you that it literally does not know what you are trying to say. What you said was "unexpected."

   ```
   >> 5a = 2
   Error: Unexpected MATLAB expression.
   ```

   And Python will tell you similarly, that it was invalid:

   ```
   >>>5a = 2
   SyntaxError: invalid syntax
   ```

   b. Having a space within a variable name. It has to be one word. Cryptic error message in MATLAB:

   ```
   >> n ame = 2
   Too many input arguments.
   ```

   And generic error message in Python:

   ```
   >> n ame = 2
   SyntaxError: invalid syntax
   ```

   c. Having a special character in a variable name. Can't be done in MATLAB:

   ```
   >> a$ = 5
   Error: The input character is not valid in MATLAB statements or expressions.
   ```

   Or Python:

   ```
   >> a$ = 5
   SyntaxError: invalid syntax
   ```

3. *Trying to access matrix elements with invalid indices:* MATLAB indexes from 1 (the first element is element 1), so indices can't be 0 or negative. And they have to be integers. Example:

```
>>a=[2,3,4];
>> a(0)
>>a(1.5)
Subscript indices must either be real positive integers or logicals.
```

In Python:

```
>>>a=[2,3,4]
>>>a[1.5]
TypeError: list indices must be integers, not float
```

4. *Trying to access matrix elements that exceed the size of the matrix:*

```
>>a=[2,3,4]
>>a(5)
Index exceeds matrix dimensions.
```

Or in Python:

```
>>> a = [3,4,5]
>>> a[6]
IndexError: list index out of range
```

5. *Trying to access a variable that doesn't exist:* This usually happens when trying to access a variable that no longer exists (has been cleared from memory), hasn't been declared yet (Matlab reads and executes lines in strictly chronological fashion) or simply when misspelling a variable that has been declared (although Matlab now ventures a guess in this case):

```
>> times5 = 5;
>> tines5
Undefined function or variable 'tines5'.
```

6. *Using a variable name that corresponds to a function, then trying to use that function:* Generally speaking, it is advisable to use descriptive variable names ("participantPoolSize" instead of "size" to avoid collisions with functions, as they are usually short). If you want to be sure whether the name of a variable you plan to use corresponds to that of an

existing function you can type `which` followed by the variable name. If nothing comes up, you are good to go. What makes this error particularly tricky is that the error message you will see depends on the particular function you replaced. The dead giveaway that you made this error is simply that a function that you used many times before is suddenly no longer working. Note that the last of these examples is the most insidious, as it doesn't even throw an error message—it dutifully does what you asked it to do (but probably not what you meant), namely to access the first element (as denoted by the content of A) of "mean"—a classic man–machine misunderstanding.

```
>> size = 7;
>> size(A)
Index exceeds matrix dimensions.
```

```
>> sin = 180;
>> sin(pi)
Subscript indices must either be real positive integers or logicals.
```

```
>> mean = 5;
>> A = 1;
>> mean(A)
ans =
5
```

7. *There can be no holes in a matrix:* **Each row and column in a matrix needs to have the same number of elements. If this isn't so, MATLAB and Python will throw an error:**

```
>> A = [1 2 3;4 5]
```
Error using vertcat

```
Dimensions of matrices being concatenated are not consistent.
```

8. *When adding matrices, make sure that they have the same dimensionality:*

```
>> A = [1 2;3 4];
>> B = [1 1 1; 2 2 2];
>> C = A + B
Error using +

Matrix dimensions must agree.
```

This might seem like a subtle point, but this also applies to adding *to* a matrix. In the commands below, assigning A to the first column of C sets C up as a 5×1 matrix, but B is only four elements long. You can't then assign B to the second column of C. The two assignment dimensions don't match, and Matlab will, helpfully if frustratingly, tell you so.

```
>> A = [1 2 3 4 5];
>> B = [6 7 8 9];
>> C(1,:) = A; C(2,:) = B;

Subscripted assignment dimension mismatch.
```

9. *When plotting vectors, the x and y must be the same length:* Generally speaking, every function that does something with pairs of vectors, the number of elements in the two vectors needs to match:

```
>> A = [1 2 3 4 5];
>> B = [6 7 8 9];
>> plot(A,B)
Error using plot
```

Vectors must be the same length.

10. *When specifying attribute-value pairs, make sure to give it what it expects:* A number if it expects a number and a string or character when a string or character is expected. Characters or strings are denoted by single scare quotes, whereas numbers lack quotation marks. All attributes of a plotted line can be specified, for instance "linewidth," which expects a number or "marker," which expects a string:

```
>> plot(A,B,'linewidth','3')

Error using plot

Value not a numeric scalar

>> plot(A,B,'marker',square)

Undefined function or variable 'square'.
```

The same goes for parsing errors (everything inside is seen as part of the attribute, and colored purple in MATLAB—i.e., the dead giveaway):

```
>> plot(A,B,'linewidth,'3)

plot(A,B,'linewidth,'3)

|

Error: Unexpected MATLAB expression.
```

11. *In Python, blank space matters:* while useful for making code readable in MATLAB, blank spaces in Python are required for any loops or function definitions. For example, inside of a for loop, you should indent with four spaces. If you do not, you will see this error:

```
for a in range(5):
print a

print a
      ^

IndentationError: expected an indented block
```

12. *Having the right packages/toolboxes.* In Python, you must explicitly import any package that is not core to Python (the common ones to import in this book are `numpy`, `scipy`, `matplotlib.pyplot`, and `scikit-learn`). In MATLAB, you need to make sure you have the right toolboxes. Many toolboxes in MATLAB can be pricey (e.g., the Statistics & Machine Learning Toolbox is $1000 as of June 2016). You will need to have this toolbox installed in order to use them.

Finally, you can see from all of this that MATLAB and Python both take you very literally. They will not "catch your drift." This is a feature, not a bug. It yields consistency. What an error message indicates is that the software does not understand your command well enough to execute it, but is aware of this fact. Again, this is a good thing. The worst mistake is the kind that is serious but goes unnoticed. And remember: The computer takes you 100% literally. It has to, because you have no common ground and it is an advantage because behavior is consistent.

Practical Considerations, Technical Issues, Tips and Tricks

PACKAGE INSTALLATION

To install a new package (e.g., the PubMed scraping package `biopython`), if you have installed anaconda, you can type:

```
conda install biopython
```

A more common approach is to use pip:

```
pip install biopython
```

PYTHON LIST COMPREHENSIONS

Python's *lists* are capable of holding data of different types in customized arrangements. This makes for a very flexible data arrangement. Lists are also handy as *comprehensions* allow the user to do things like *for loops* and *if statements* in single lines of code. Here are a few examples:

To unpack a list of lists, a, where:

```
>>>a = [[1,2,3],[4,5,6],[7,8,9]]
>>>unpacked = [col for row in a for col in row]
>>>print unpacked
[1,2,3,4,5,6,7,8,9]
```

To unpack the list of lists, a, if the element is greater than 5:

```
>>>unpacked_conditional= [col for row in a for col in row if col> 5]
>>>print unpacked_conditional
[6,7,8,9]
```

To regroup columns of lists into rows, for list b, the function zip(*b) returns a list of tuples of the rearranged data:

```
>>>b=[[1,2],[3,4],[5,6],[7,8]]
>>>zip(*b)
[(1, 3, 5, 7), (2, 4, 6, 8)]
```

PYTHON LISTS VERSUS NUMPY ARRAYS

Lists in Python can contain different types of data. The first element can be an integer, and the second element can be a string. Numpy arrays are built to contain an even wider variety of numerical data types than Python, including types: `bool_`, `int_`, `intc`, `intp`, `int8`, `int16`, `int32`, `int64`, `uint8`, `uint16`, `uint32`, `uint64`, `float_`, `float16`, `float32`, `float64`, `short`, `long`, `longlong` `complex`, `complex64`, `complex128`. Array manipulations in numpy are often faster than list comprehensions in Python as many numpy packages are written in C (under the hood, i.e., so no need to learn C). See the numpy documentation online for full descriptions.

TEXT EDITORS, THE COMMAND LINE, HOW TO GO BETWEEN SUBLIME AND THE TERMINAL

When using Python, there are many ways to create code and visualize data. One common way is to use a text editor (such as Sublime Text, which shows the Python code on the right) and save the text file (shown as `easy_example.py`). You can then run the code from a Terminal, typing `python easy_example.py`, and the code will run (and the figures will be displayed if you use the command `plt.show()`). Make sure your working directory (in the terminal) is in the same folder that you saved the file `easy_example.py` (Fig. C.1).

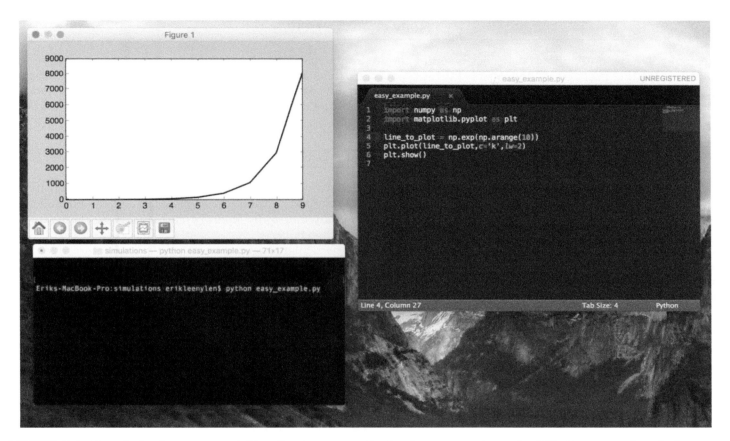

FIGURE C.1 Example development setup: Sublime Text (or any text editor) and save your code as a ".py" file. You can run the code from your terminal.

PYTHON ON WINDOWS

A common issue on Windows machines is setting the environmental path—this is how the compiler knows where to look when loading packages. If you create your own packages, functions, or classes, or if you download external packages manually into folders, then you will need to add these to the path. An easy trick for this on Windows is to select: *Win+Pause*, then select the *Advanced,* and select *Environment Variables*. In the *Edit System Variable*, you can add folders to this list, separated by a semicolon (;).

JUPYTER: USING IT AND ITS GREAT FUNCTIONS

The Jupyter notebook is a great friend of the data scientist. It allows the user to write code and create visualizations of data all in the same tab on their browser. It is included in the standard distribution of *Anaconda*, and can be launched from the command line (note, not inside Python, but in the *terminal* window) by entering *jupyter notebook*. On the upper right side, select *New*, and then select *Python*. This will open a new notebook.

The *matplotlib inline* feature uses the magic operator %. This particular magic operator prints graphs directly in the notebook. This is perhaps the most friendly feature to any scientist: code followed by visualization, in one neat notebook. This allows for effective notebooks to be extremely shareable. Note that to execute a block of code you can type shift+enter.

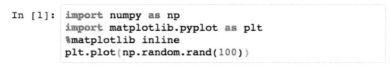

```
In [1]:  import numpy as np
         import matplotlib.pyplot as plt
         %matplotlib inline
         plt.plot(np.random.rand(100))
```

Out[1]: [<matplotlib.lines.Line2D at 0x109f136d0>]

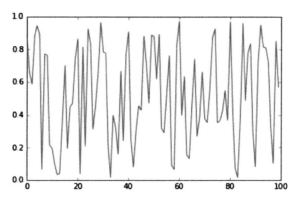

One advantage to Jupyter is its built-in *help* function. To see the documentation for a particular function, try, e.g.:

```
>>> import numpy
>>> help(numpy.random.randint)

Help on built-in function randint:

randint(...)
    randint(low, high=None, size=None)
    Return random integers from `low` (inclusive) to `high` (exclusive).
    Return random integers from the "discrete uniform" distribution in the
    "half-open" interval [`low`, `high`). If `high` is None (the default),
    then results are from [0, `low`).

    Parameters
    ----------
    low : int
        Lowest (signed) integer to be drawn from the distribution (unless
        ``high=None``, in which case this parameter is the *highest* such
        integer).

    high : int, optional
        If provided, one above the largest (signed) integer to be drawn
        from the distribution (see above for behavior if ``high=None``).
    size : int or tuple of ints, optional
        Output shape. If the given shape is, e.g., ``(m, n, k)``, then
        ``m * n * k`` samples are drawn. Default is None, in which case a
        single value is returned.

    Returns
    -------
    out : int or ndarray of ints
        `size`-shaped array of random integers from the appropriate
        distribution, or a single such random int if `size` not provided.

    See Also
    --------
    random.random_integers : similar to `randint`, only for the closed
        interval [`low`, `high`], and 1 is the lowest value if `high` is
        omitted. In particular, this other one is the one to use to generate
        uniformly distributed discrete non-integers.
```

```
Examples
--------
>>> np.random.randint(2, size=10)
array([1, 0, 0, 0, 1, 1, 0, 0, 1, 0])
>>> np.random.randint(1, size=10)
array([0, 0, 0, 0, 0, 0, 0, 0, 0, 0])

Generate a 2 x 4 array of ints between 0 and 4, inclusive:

>>> np.random.randint(5, size=(2, 4))
array([[4, 0, 2, 1],
   [3, 2, 2, 0]])
```

Jupyter notebooks (soon to be released as *Jupyter Lab*) are also great tools for collaboration as they can be shared and rendered on *GitHub*. This allows for your pretty visualizations and your code to be shared on the web.

THE BIGGEST DIFFERENCES BETWEEN PYTHON 2 AND 3

1. Division. Python 3 does floating point division between two integers if you specify the division with two slashes (//). However, if we want Python 2 to do floating point division by default (as opposed to integer division as seen in chapter: From 0 to 0.01), we can use the handy function:

```
>>> from__future__ import division
>>> 3/2
1.5
```

2. Printing. In Python 3, you have to enclose any *print* statements in parentheses, e.g., *print('hello')* in Python 3 instead of *print 'hello'* in Python 2.
3. Range. In Python 3, the *range* function is an iterable object. We didn't really go into what this means in this book. But note that it won't return a *list* if you try to print it.

Obviously there are a bunch more differences between Python 2 and 3, most of which are beyond the scope of this book. Note that differences in Python 3 have resulted in significantly faster code compared to Python 2.

CONVENTIONS IN PYTHON

Standard Python convention uses lowercase letters for the first letter of variable names when describing lists, arrays, dictionaries, and *methods*, but uses capital letters at the beginning of *Classes*. However, you can legally use any combination of upper and lowercase letters in naming variables and your computer won't care (however, your [future] employer might care very much).

MATLAB TIPS AND TRICKS

Version Issues

One of the key issues to be mindful of is that there are two new versions of MATLAB every year, one in the spring of that year (e.g., 2016a) and one in the fall (e.g., 2016b). Whereas changes in each release are incremental, over time, these changes do add up to transformational change. All MATLAB code in this book should work with the 2016b release—if something doesn't run as it should and you are currently using a newer version of MATLAB, check the release notes on the Mathworks website. For instance, between MATLAB 2014a and now, the default colormap was changed from "jet" to "parula." If this changes again in the future, the figures in the book might look different from what you see. If that is the case, you can set your colormap explicitly, by typing colormap(jet) or colormap(parula). Sometimes, changes to how MATLAB works require updating your code. For instance, the way to initialize ("seed") the random number generator was by calling the setDefaultStream method of the RandStream object, such as this: RandStream.setDefaultStream(s);
after defining s as s = RandStream('mt19937ar', 'seed', sum(100*clock))
This specifies a particular type of random number generation method (one involving mersenne twisters), hooked up to the system clock.

But this no longer works. The method setDefaultStream has been replaced with setGlobalStream, the correct code is now: RandStream.setGlobalStream(s);

Be on the lookout for things like that.

VECTORIZATION

MATLAB is an interpreted language, so each line is interpreted and executed one after the other. This is fine, but if there are a lot of lines, it can take a lot of time to execute the code. This is a particular concern if there are long loops, and possibly even nested loops. In principle, every loop can be replaced by a vector operation, and MATLAB is optimized to do those, so this will speed up your code considerably. Here, we will provide three simple examples of how to do this that generalize easily. Note that this is less of a concern as of late. Which interfaces with the "version issues" point made above. Since recently, Matlab code is now auto-compiled before it is run (under the hood and out of sight), speeding up code considerably.

1. A single loop:

Say you have data from 100,000,000 trials and need to calculate the total number of photons presented in a given trial (from illumination levels and time presented in milliseconds). If you do this with a loop, it will take a while:

```
numTrials = 1e6;
numPhotons = randi(100,[numTrials,1]);
exposureDuration = randi(1000,[numTrials,1]);
tic
for ii = 1:numTrials
   totalExposure(ii,1) = numPhotons(ii)*exposureDuration(ii);
end
toc
```

If you replace the second paragraph with this one (without the loop) the result will be the same, but it will be much faster:

```
tic
totalExposure = numPhotons.*exposureDuration;
toc
```

2. Nested loops:

Say you have a 10,000 by 10,000 matrix that results from multiplying all numbers from 1 to 10,000 with all other numbers from 1 to 10,000 (a full cross). You can do this element by element, first going through all rows, then all columns. Note that we always preallocate. Otherwise, this would take even longer:

```
howBig = 1e4;
tic
M = zeros(howBig,howBig); %Always preallocate
   for ii = 1:howBig
      for jj = 1:howBig
         M(ii,jj) = ii*jj; %Each ii, jjth entry of M is ii * jj
      end
   end
toc
```

This works, but it takes a long time.

Now, we vectorize the last dimension (columns), so instead of a nested loop, we have a single loop:

```
tic
M = zeros(howBig,howBig); %Always preallocate
for ii = 1:howBig
   M(ii,:) = ii*(1:howBig); %Doing each row at once
end
toc
```

This should already be much faster.

Finally, let's vectorize both dimension and get rid of loops altogether:

```
tic
M = zeros(howBig,howBig); %Always preallocate
M(:,:) = (1:howBig)'*(1:howBig); %Doing it all at once
toc
```

Note that we have to transpose the first vector to get the outer product. All three code versions yield the same result, but you should be able to realize considerable time savings with the last one. These would be even more dramatic if you compared it to a unpreallocated version of the code.

3. Conditionals:

Say you want to add a number (e.g., performance) to a running total, but only if another number (e.g., percentage of trials completed) is big enough. You could do this in a loop, checking the condition each time.

```
numParticipants = 1e6;
numTrials = randi(100,[numParticipants,1]);
performance = rand(numParticipants,1);
cumPerf = 0;
tic
for ii = 1:numParticipants
   if numTrials(ii,1) > 50
      cumPerf = cumPerf + performance(ii,1);
   end
end
toc
```

It is straightforward to replace the second paragraph with faster code that produces the same result and that gets rid of the loop:

```
tic
temp = find(numTrials > 50);
cumPerf = sum(performance(temp));
toc
```

PRACTICAL CONSIDERATIONS

Highly subjective ranking of marketable attributes when looking for a job

	Neuroscience	Data Science
1.	Publication list (peer-reviewed)	Programming skills (GitHub repo)
2.	Grants/awards	Communication skills (interview)
3.	Pedigree (mentors/universities)	Machine learning skills (courses, competitions)
4.	Teaching	Job experience (resume)
5.	Letters of recommendation	Problem-solving abilities (white board)
6.	Interview	References/network

Neuroscience interview preparation checklist:

1. Study the papers of the interviewer
2. Understand what the interviewer's lab does
3. Dress for the job you want
4. Practice interviewing (have someone else quiz you) and be able to describe why you want to join that institute
5. Have questions ready for the interviewer
6. Practice a 10-minute description of your previous work
7. General study
   a. Read *Principles of Neural Science* (Kandel and Schwartz)
   b. Read *Spikes* (Rieke et al.)
   c. Read this book
   d. Be able to describe a realistic experimental design in 5 minutes

    e. Address two hypotheses you'd like to ask
    f. Know the latest methods in your particular field
    g. Know some Python
    h. Know some MATLAB
    i. Know some statistics

Data science interview preparation checklist:

1. Study the company
2. Understand how the company makes money
3. Dress slightly above how you expect the company dresses
4. Practice interviewing (have someone else quiz you) and be able to answer why you would be a good fit at that company
5. Have questions ready for the interviewer
6. Practice a 2-minute description of your previous work
7. Have programs in each language you are proficient in on your Github page
8. General study
    a. Sorting algorithms
    b. Know Python
    c. Web-based applications
    d. Know your tech lingo
    e. Be ready for a data challenge
    f. Do probability problems on a white board
    g. Know regularization
    h. Understand crossvalidation issues
    i. Be able to wrangle and model some data
    j. Know about optimization
    k. Know some SQL
    l. Study data structures in computer science
    m. Study recursion in computer science
    n. Know AWS
    o. Know general Business Intelligence (BI) terminology
    p. Understand A/B testing
    q. Know some statistics
    r. Know some machine learning

Glossary (Including Additional Python and MATLAB Packages and Examples)

	Type: Python, MATLAB, Neuroscience, Statistics, Data Science, Computing, Engineering, Math, Physics, Life, Medicine, Biology, Psychology, Machine Learning, Industry	Definition
5HT-3A	Neuroscience	5HT is serotonin, a monoamine neurotransmitter, thought to be important for mood. 3A is one of its receptor subtypes
Acetylcholine	Neuroscience	ACh, a neurotransmitter involved with muscular control, attention, learning and other brain mechanisms
ACSF	Neuroscience	Artificial cerebro-spinal fluid, a fluid that mimics that in which the brain resides, useful for slice (in vitro) preparations of brains
Action potential	Neuroscience	A voltage spike traveling down an axon, where the neuron sends a signal to one or more neurons. Usually conceptualized as a binary event
Adaptation	Neuroscience	The process by which neural responses (spikes, graded potentials, behavior) change their output in response to stimulation
Afferent	Neuroscience	Signals (typically action potentials) that are traveling towards to the brain. One good example is the optic nerve: all action potentials travel from the retina towards the brain (specifically the waystation of the lateral geniculate nucleus of the thalamus)
After-hyperpolarization	Neuroscience	A downward deflection (that is, more negative) of a neuron's voltage that occurs after spikes
Agile method	Computing	Method for software development in teams, used for fast and adaptive creation of code
Aliasing	Engineering	An effect where sampling of a signal at too low a frequency results in the appearance of another, artifactual signal

(Continued)

	Type	Definition
Alpha wave	Neuroscience	EEG waves with a frequency between 7.5 and 12.5 Hz—usually associated with relaxation
Alzheimer's disease	Neuroscience	Senile dementia; degenerative disease associated with loss of neurons, memory as well as personality and behavioral changes
Amacrine cell	Neuroscience	Class of retinal interneurons thought to modulate signal transmission in the retina
AMPA receptor	Neuroscience	Ionotropic receptor, commonly associated with excitatory glutamatergic transmission in neurons
Amplifier	Engineering	apparatus for making a signal stronger; something that amplifies
Amygdala	Neuroscience	Almond-shaped structure in the brain, thought to be important for processing fear, emotion, and threats. Part of the limbic system
Anesthesia	Neuroscience	A state of induced lack of consciousness, typically induced by anesthetic agents for the purposes of analgesia during surgery
Aneurysm	Neuroscience	Ballooning of an artery, can cause serious damage if it occurs in the brain
Apparatus (plural: apparatus)	Neuroscience, Engineering, Life	Thing, tool, instrument
Aperiodic	Math	Not occurring regularly; not periodic
Apical dendrite	Neuroscience	Dendritic processes of a neuron that typically project "up," in cortical neurons these project towards the cortical surface (see also *dendrite*)
Apoptosis	Neuroscience, biology	The (active) process of cell death
append	Python (numpy)	Python function for adding a value to the end of an array
Array	Python, math	An ordered list of numbers (Python); a multidimensional matrix (Math)
Artificial intelligence	Neuroscience, computing, data science, machine learning	Machines that are smart. In contrast to natural stupidity
Astrocyte	Neuroscience	Cells in the CNS and PNS thought to serve as caretakers of neurons. In charge of logistics and immune functions
Auditory nerve	Neuroscience	Afferent nerve fiber carrying electrical information from the cochlea to the brainstem

(Continued)

	Type	Definition
Autism	Neuroscience, medicine	Spectrum of disorders in humans characterized by problems with communication and social interactions as well as stereotyped behaviors
AWS	CS	Amazon Web Services: a range of cloud computing tools
axhline	Python (matplotlib.pyplot)	Python function for plotting horizontal line
Axon	Neuroscience	The projection of a neuron on which the action potential travels to synapse with another neuron
Axon hillock	Neuroscience	Point of spike initiation in a neuron, an action potential is triggered to travel down an axon once membrane voltage exceeds the spike threshold
axvline	Python (matplotlib.pyplot)	Python function for plotting vertical line
Basemap	Python, data science	Python library for using geographic data
Beta wave	Neuroscience	EEG waves with a frequency between 12.5 and 30 Hz—usually associated with consciousness, usually the default waking state
Bimodal distribution	Stats	Distribution that appears to have two humps
Bipolar cell	Neuroscience	Cell type in the retina, typically receiving inputs from photoreceptors and sending outputs to amacrine cells and ganglion cells
Bits	Computing	Binary digit (0 or 1)—there are 8 bits in a byte. Conceived of by John Tukey
Blindness	Medicine	The lack of vision
Bokeh	Python	Python library for visualization
Bootstrap	Statistics	Technique in statistics for resampling data, typically used in cases of low sample numbers or for class imbalances (when two sample sizes are different, and therefore imbalanced). Introduced by Bradley Efron
Borosilicate glass electrode	Neuroscience	Type of electrode used for patch clamp recordings
Boyden, Ed	Neuroscience	Pioneer in the field of optogenetics
Brain	Neuroscience, medicine, life	Principal structure for animal behavior, perception, computation, decisions, personality, consciousness, creativity, and war

(Continued)

	Type	Definition
Brain–machine interface	Neuroscience	Tool for interacting electronically with a brain in either direction (world to brain or brain to world)
Brainstem	Neuroscience	Structure that acts as the gateway between the spinal cord and the brain. Houses many critical systems that regulate waking and sleeping, breathing, etc.
Burst	Neuroscience	The fast firing of two or more action potentials in quick succession
Butterworth filter	Engineering	Filter characterized by a flat passband
Cable theory	Neuroscience	Collection of models used to describe the flow of electrical activity in passive neuronal processes. Originally conceived to describe electricity flow in underwater telegraph cables across the Atlantic
Cajal, Santiago Ramon	Neuroscience	Nobel Prize-winning neuroanatomist, mostly known for drawing detailed images of the brain using microscopy and the neuron doctrine—the notion that discrete neurons are the fundamental computational unit in the brain
Calcium	Neuroscience	Ion, important for many intracellular signaling cascades
Callosotomy	Neuroscience, medicine	Removal of the corpus callosum
Callosum	Neuroscience, medicine	The information highway containing large bundles of axons between the two hemispheres of the brain
Capacitance	Engineering/Physics/ Neuroscience	The characteristic ability to store charge. Property of cell membranes and batteries
Capacitor	Engineering/Physics	Thing that stores an electric charge
Cartesian coordinates	Math	The x and y axes. Introduced by Rene Descartes
Cassandra	Computing	Type of database pioneered by Facebook
Cell-attached recording	Neuroscience	Method for recording the membrane voltage of a cell, useful for recording spikes from single cells
Cerebellum	Neuroscience	Latin for little brain, highly complex structure with nearly as many cells as the rest of the brain (in humans), important for movement, timing, and the execution of highly practiced movement sequences, among other things

(Continued)

	Type	Definition
Cerebrospinal fluid	Neuroscience, medicine	See CSF
Channel	Neuroscience/Biology	Membrane protein that can actively or passively allow for the flow of ions in or out of the cell.
Channel 2 rhodopsin	Neuroscience	Channel popularized for being activated by light
Chebyshev filter	Engineering	Type of filter with a very steep roll off
Class	Computing	A programmatic template for creating objects
Classification	Machine learning	Type of supervised learning
Cloud	Computing	Notion of performing computations or storing data on a separate, often times larger and more powerful, device, usually connected via the internet
Clustering	ML	Type of unsupervised learning for predicting how data group together
Cochlea	Neuroscience	Organ that transduces physical waves (sounds) into electrical signals for further processing in the brain
Code-safe	Neural data science	Inspired by "swim-safe," where a person is trained to simply not drown in water, code-safe implies that one knows enough to not drown in the data
Collaborative Filtering	Machine Learning	Type of recommendation engine
collections	Python	Package in Python with several high-performance container datatypes, notably *defaultdict* in this book. Also includes *deque*, *Counter*, *OrderedDict*
Comment	Python, MATLAB	Method for annotating computer code that the machine ignores but are readable by humans
Complex conjugate	Math	A number whose imaginary component has the opposite sign
Conductance	Engineering, neuroscience	How much something conducts electricity; reciprocal of resistance
Cone	Neuroscience	Photoreceptor. Thought to underlie color vision and high acuity daylight vision. Most common in the fovea. In humans, there are three types
Confidence interval	Statistics	Interval in which a population parameter (usually the mean) lies with a certain probability

(Continued)

	Type	Definition
Connectome	Neuroscience	A map of the connections between all neurons in the brain
Constructive Interference	Engineering	The union of two waves such that the peaks align, yielding a wave with larger amplitude
Correlation	Neuroscience, statistics	Metric for quantitatively evaluating the relationship between two variables
Cortex	Neuroscience	The evolutionarily recent neural structure that makes up the wrinkly lobes in mammals, often thought to be the most complex system we've encountered (so far) in the universe; what makes humans human. Covers the rest of the brain (in humans)
CPU	Computing	Central processing unit; typically a microprocessor that acts to carry out programmatic commands for a computer
Craniotomy	Neuroscience	Surgical method of removing part of the skull for access to the brain
Crossvalidation	Machine learning	Method of implementing multiple training and testing sets in machine learning to detect and prevent overfitting
CSF	Neuroscience	Cerebrospinal fluid, acts as nourishment and projective matter for the brain
Current clamp	Neuroscience, engineering	Method for holding the current at a fixed value to as to record the voltage
Data	Data, computing, neuroscience, life	The quantified result of a systematic measurement process, capturing information about the natural world
Data science	Data science	Field dedicated to understanding the structure of data itself, regardless of field
Deafness	Medicine, neuroscience	Lack of hearing
Decision making	Neuroscience, life	The process of taking inputs to calculate a desired or optimal output
Decoding	Math, algorithms, statistics, neuroscience	The conversion of information into meaningful output
Decorator	Python	Tool for "wrapping" function in other functions
Deep learning	Machine learning	Method in machine learning, using multilayer neural networks to create predictive models that are often opaque even to the people that created them

(*Continued*)

	Type	Definition
def	Python	Name for declaring a method or function
Deisseroth, Karl	Neuroscience	American neuroscientist credited with pioneering methods in optogenetics
Dendrite	Neuroscience	Processes of a neuron that receive inputs from other neurons. Dendrites, named after their tree-like appearance (dendros = Greek for tree), can make synapses with axon terminals from other neurons
Desensitization	Medicine, neuroscience	The loss of sensitivity
Destructive Interference	Engineering	The union of two waves such that the peaks do not align, yielding a wave with smaller amplitude
Dictionary	Python	Object containing key:value pairs
Dimensionality reduction	Math	Method for reducing data in many dimensions into a lower-dimensional space to try to make sense of things
DNA	Biology/Life	Deoxyribonucleic acid. The genetic roadmap for the architecture of life. Typically inherited from parents.
Dopamine	Neuroscience	Monoamine neurotransmitter important for reward signaling, motivation and movement
Durotomy	Neuroscience	Surgical removal of the dura mater
e	Math	Euler's number, close to 2.71
Edge artifacts	Engineering	The result of applying a filter to an edge of data, commonly observed when a filter is applied at time = 0
Edge case	CS, industry	Situations that appear uncommonly but must be considered in advance and can often result in serious code bugs, such as unforeseen changes in array size
Efferent	Neuroscience	Any projection transmitting information away from the brain (opposite of afferent)
Eigenvalue	Math	The length of an eigenvector
Eigenvector	Math	Linear operations don't change the direction of the eigenvector, they only scale it. Think of all points that connect the poles on earth. They lie on the eigenvector, as the rotation of the earth doesn't change their position

(*Continued*)

	Type	**Definition**
Electrode	Neuroscience	Apparatus for recording electrical activity (in neuroscience usually from the brain)
Electroencephalography	Neuroscience	EEG, method for recording electrical activity from the scalp, popular in the study of sleep and attention, introduced by Hans Berger
Electrophysiology	Neuroscience	Field of studying the electrical properties of physiological systems
Encoding	Neuroscience, Math	Conversion of stimuli into meaningful information in the brain
Engram	Neuroscience	Where and how memories are stored in the brain
Ensemble	Neuroscience, Machine Learning	Many neurons at once (Neuroscience); the use of many techniques to make the best prediction (ML)
Entropy	Neuroscience, physics	Measurement of disorder or information in a signal. Basically quantified surprise
Epigenetic	Biology	Pertains to switching genes on and off, and gene expression more generally. Can potentially account for passing acquired traits to future generations, e.g. the effects of famine on offspring
EPSP	Neuroscience	Excitatory post-synaptic potential. A voltage deflection in a target neuron (post-synapse meaning after the synapse, or on the receiving end) that causes the voltage to move towards the cell's spike threshold, typically increasing the probability of firing.
ETL	Data science, industry	Extract, Transform, and Load, method for data transformation
Euler's formula	Math	$e^{ix} = \cos(x) + i{*}\sin(x)$
Excitation	Neuroscience, biology	Typically refers to the depolarization of a cell towards or past its spike threshold
Experiment	Life	Experiments systematically create or manipulate conditions of interest while keeping everything else constant, usually by randomization. Potentially allows to uncover causal links between variables
Factorization	Math	Method for decomposing an array
Fast Fourier transform	Engineering	Method for the fast digital implementation of the Fourier transform, introduced by Tukey

(Continued)

	Type	Definition
Fear conditioning	Neuroscience, psychology	Use of aversive stimuli to study how threats are perceived and remembered (also: threat conditioning)
Feature engineering	ML	Selection of features on which to perform analysis
Feed-back	Engineering, neuroscience	System in which outputs of a structure can project back to earlier inputs
Feed-forward	Engineering, neuroscience	System in which serial input–output structures are unidirectional
FEF	Medicine, neuroscience	Frontal eye field; part of frontal cortex
FFT	Engineering	See Fast Fourier Transform
Filter	Engineering	An array used for transforming data in a certain way
Filtering	Engineering	To apply a filter, typically to a signal, or coffee
Fitting	Statistics/ML	Figuring out what function (often a line) best approximates the data
fMRI	Neuroscience	Functional magnetic resonance imaging, often in living brains, usually relying on the hemodynamic response
Fortran	CS	Old, fast programming language
Fourier transform	Engineering	Method for converting signals in the time domain to signals in the frequency domain
Fovea	Medicine, neuroscience	Part of the retina where cone photoreceptors exist in high density, critical in humans for perception of visual elements that require high acuity
GABA	Neuroscience	Gamma-aminobutyric acid, a principal inhibitory neurotransmitter in the brain
Gain	Engineering	In the analogy of a volume on a stereo: How turned the knob is
Gap junction	Neuroscience, biology	"Electrical synapse," allows for the exchange of electrical signals between neurons
Gaussian	Stats	Also known as normal, a type of distribution commonly known as a bell curve
GFP	Neuroscience	Green fluorescent protein, used for tagging and visualizing cells
Git	Computing	Version control software
Github	Computing, Industry	Popular website for version controlling software

(Continued)

	Type	Definition
Glia	Neuroscience	Helper cells in the brain, provide logistical and structural support to neurons
Glutamate	Neuroscience	Principal excitatory neurotransmitter in the brain
GPU	Computing	Graphics processing unit, computing machinery designed for rapid visualization
Grating	Neuroscience, psychology	Common type of visual stimulus containing oriented stripes. Popular because it can be completely defined by a few parameters such as location, size, contrast, orientation, and spatial frequency as well as the type of grating (e.g., sinusoidal vs. bar, indicating how contrast is modulated)
Gray matter	Neuroscience, medicine	Large-scale coloring of cell somata (in contrast to white matter)
Grid cell	Neuroscience	Cell in the hippocampus involved with processing the location of one's body in space. Akin to a neural GPS
Grid search	Python	Scikit-learn tool for searching for optimal parameters in a model
Ground	Engineering	Electrical connection to earth, or to a reference point, e.g., in the EEG, usually the ear
Ground truth	Neuroscience, Statistics, Machine Learning, Life	Veridical information about a system, known to the researcher. Useful to train an algorithm or to assess the accuracy of a model or simulation
Gyrus	Neuroscience	Cortical fold, "hill"
h-index	Neuroscience	Contentious metric for quantifying how good someone's publication record is. If someone's h-index is 5, they have to have five publications that each have been cited five or more times
Hadoop	Computing, data science	Open source software framework for distributed processing and data storage. Particularly useful for large datasets
Hanning Window	Engineering	Window shape with a particular side-lobe roll off characteristics
Harmonic oscillators	Neuroscience	Things that don't exist in the brain on the single cell level
Hebb's rule	Neuroscience	Cells that fire together wire together
HFT (high-frequency trading)	Data science	Method for rapidly buying and selling financial data or property

(*Continued*)

	Type	Definition
High Gamma	Neuroscience	Neural oscillations or signals in the 60-200Hz range, sometimes associated with spiking activity
Hillock	Neuroscience	See axon hillock
Hippocampus	Neuroscience	Structure in the brain important for memory consolidation and spatial navigation
Hyperacusis	Neuroscience/medicine	Sensitivity to sounds where certain stimuli are painful to the observer
Hypothesis	Science	Theory, guess, something that should be tested with data
i	Mathematics	Square root of minus one, an imaginary number
IC	Neuroscience	See Inferior Colliculus
ICA	Stats	Independent component analysis, method for dimensionality reduction and source separation
Impact factor	Neuroscience	Metric for assessing citations that a journal receives
In silico	Computing, neuroscience	Literally in silicon; on a chip; simulation
In vitro	Neuroscience/Biology	In a dish; common name for methods that use thin slices of brains
In vivo	Neuroscience/Biology	Live; condition for studying an animal that is alive
Index	CS/Python/MATLAB	The relative numerical position of an element
Inference	Stats	Using reasoning to determine a conclusion
Inferior Colliculus	Neuroscience	Midbrain structure important for relaying auditory information from brainstem structures to the medial geniculate body (MGB) of the thalamus. Also receives considerable multi-sensory input and cortical feedback
Information theory	Engineering/data science	Field pioneered by Claude Shannon; quantification of information
Inheritance	CS	Object property of inheriting properties of other objects. A biological analogy would be a cell inheriting the genetic information of its mitochondria.
Interneuron	Neuroscience	Technically speaking, any neuron contained within a small region (not projecting to other areas); colloquially, inhibitory cells whose activity suppresses the activity of other cells

(*Continued*)

	Type	**Definition**
Invasive	Neuroscience	Requiring perturbation, pain, or surgery
Inverse Fourier transform	Engineering	Takes the signal from the frequency domain back to the time domain
Ionotropic	Neuroscience	Type of fast membrane channel
Ions	Physics	Charged elements or molecules
IPSP	Neuroscience	Inhibitory postsynaptic potential
iPython	Python	Command shell for interactive computing in Python
ISI	Neuroscience	Interspike interval, time between spikes
Javascript	Computing	Programming language, the "language of the internet" (as of 2016)
JSON	Computing, data science	Javascript object notation, common data structure type
Julia	Computing	High-level, high-powered fourth-generation computing language, inherently supporting multiple processes
Jupyter	Python/Computing	Brand of Python notebook, useful for the analysis and visualization of data
K-nearest neighbor	Machine Learning	Machine learning method for classification and regression, where data are classified according to a generalized measure of what groups their neighbors belong to
Kaiser window	Engineering	A family of windows in digital signal processing
Kaiser window	Engineering	A window function used in signal processing
Kernel	Engineering, Math, Computing, ML	The core unit of interaction; a filter (engineering); the central core of a computer's operating system (CS); method for using customized or specialized nonlinear classification parameters (ML, as in SVM)
KMeans	ML	Machine learning method for clustering
Latency	Neuroscience, engineering	Time that has passed relative to some starting point
Lateral Geniculate Nucleus	Neuroscience	Principal target of the optic nerve, projecting to primarily the primary visual cortex

(Continued)

	Type	**Definition**
Layers (cortex)	Neuroscience	Organization of cortex into sheet-like arrays of similar cell types. Most mammalian cortices are sub-divided into six layers, often with sub-layers (for example, layer 4 of monkey visual cortex can be divided into 4A, 4B, 4C, which can be sub-sub-divided into 4Calpha and 4Cbeta). Also known as laminae.
Leak Conductance	Neuroscience	A measure of how permeable ions are about the membrane while a cell is in its resting state
Least squares regression	Statistics	Method for finding the line (or hyperplane) that minimizes the sum of the squared distance between the line and points
LFP	Neuroscience	See local field potential
LGN	Neuroscience	See Lateral Geniculate Nucleus
Linear regression	Statistics, machine learning	Method for modeling how well the relationship between two variables can be described by a straight line
Linearity	Statistics, machine learning	Relationship described by a straight line; a system that fulfills homogeneity and additivity
List	Python	Canonical object type in Python, known for its powerful comprehensions. One example list comprehension to unpack a nested list of lists (called nested_list) is: [b for a in nested_list for b in a]
Loadings	Mathematics/Engineering	Magnitude of coefficients, often in PCA
Lobotomy	Neuroscience	Surgically disconnecting the prefrontal cortex from the rest of the brain
Local field potential	Neuroscience	Aggregate signal from activity of many neurons, often thought to be the result of synchronized denritic activity
Logistic regression	Statistics, Machine Learning	Method for creating a model that can predict the probability of a categorical output
Luftschloss	German	Wishful thinking bouncing off the echo chamber of your own imagination; willful pipe dream
Mathematica	Computing, neuroscience	Software for mathematical analysis, introduced by Stephen Wolfram
MATLAB	MATLAB	Matrix laboratory. Software for engineering and scientific computing

(Continued)

	Type	**Definition**
Matplotlib	Python	Function for visualizations in Python, similar to many MATLAB functions
Matrix	Math, Python, MATLAB	Typically a two-dimensional array. In general, an ordered arrangement of numbers in multiple dimensions
Medial superior olive	Neuroscience	Brainstem structure important for the propagation of auditory information (especially related to the azimuth of sound) from the cochlear nuclei to the inferior colliculus
Membrane	Neuroscience	The lipid bilayer that forms the outer boundary of neurons
Membrane capacitance	Neuroscience	The amount of charge that a membrane can hold
Memory	Computing, neuroscience	The storage of data or information
Metabotropic	Neuroscience	Type of receptor characterized by indirect activation of channels
Method	Python	A function that is a member of a class. Oftentimes used colloquially to mean "function," methods are inherited properties when an object instance is called
Mitochondria	Biology	Considered the powerplants of cells, these structures give energy to cells. Originally an independent single-cell organism, still has its own DNA
Model	Stats/ML/Data Science	An equation or a system of equations. Can be useful for simplifying complicated and noisy processes to their underlying patterns, and dangerous if trusted too deeply (see Housing Bubble). When it comes to models, rent - don't buy.
Modeling	Statistics, ML	To create or implement a model, see *model*
Module	Python	A file containing Python functions, definitions, and statements. For example, you may write a module with several functions, and later import that module into another Python script
MongoDB	Industry, data science	NoSQL type database
Morlet wavelet	Engineering	A wavelet that is the result of a Gaussian multiplied with an exponential. Also called "Gabor"
Motor cortex	Neuroscience	Region of cortex responsible for coordinating, planning, and executing movement commands in the brain
MRI	Neuroscience	Magnetic resonance imaging, popular method for visualizing body tissues. Based on alignment of proton spins by magnets

(Continued)

	Type	Definition
MSO	Neuroscience	See medial superior olive
MT	Neuroscience	Middle Temporal; visual area in the primate, most closely associated with the processing of visual motion
Mus musculus	Neuroscience	Mouse
Musicophilia	Neuroscience	The admiration, love, or attraction to music; book by Oliver Sacks
Nano	CS, Physics	Text editor (CS); 10E-9 (physics)
Neo	Life	Lead character in The Matrix film trilogy
Neocortex	Neuroscience	The wrinkled, evolutionarily recent structure in mammals on top of the brain
Network	Python, math	Python package for graph analysis
Neural code	Neuroscience	Actively researched and sought language that neurons use to communicate with one another, common theories include rate, temporal, and correlation codes; the holy grail of neuroscience
Neural data science	Neuroscience, data science	The use of data wrangling, hacking, computations, and machine learning to understand the brain
Neural network	Machine learning	Series of "connected" arrays in which the weights of the connections can be trained such that output predictions can be made. Ironically named, as no actual neurons are involved and they are similar to real neurons only in the most basic ways
Neuron	Neuroscience	Electrically active cell that transmits information. Usually in the brain
Neurotransmitter	Neuroscience	Molecule or ion used for transmitting information, the most common excitatory being glutamate, the most common inhibitory being GABA
Noise	Neuroscience, engineering	Broadly defined as any part of a measured signal that is interfering with the signal of interest
Noise correlation	Neuroscience, statistics	A measure of how well two signals are correlated in the absence of a signal, often used to determine neuronal connectivity
Noninvasive	Neuroscience	Class of methods for data recording that do not require surgery

(Continued)

	Type	**Definition**
Nonlinearity	Math	Anything that is not linear
Nootropic	Neuroscience	Class of mind-altering drugs, debatably thought to enhance aspects of cognition
Normalization	Neuroscience	Method for numerical or biological scaling
NoSQL	Industry, data science	Database type with nonrelational structure (as compared to relational database syntaxes, like SQL)
Nucleus	Neuroscience	Contains most of the DNA of a cell
Nucleus accumbens	Neuroscience	Subcortical structure thought to be involved with reward processing
Numpy	Python	Python's package for numerical analysis
Nyquist frequency	Engineering	The minimum sampling rate required to prevent aliasing of a signal, twice the highest frequency component of the signal
Object	Python, MATLAB, CS	Instance of a class
OCaml	CS	Object-oriented programming language
Occipital lobe	Neuroscience	Large posterior region of neocortex, home of the visual cortices
ODE	Math	Ordinary differential equation
Ohm's law	Physics, engineering	Voltage equals current times resistance
Olfactory nerve	Neuroscience	The first cranial nerve (CN1), sends smell signals to the brain
One photon microscopy	Engineering	AKA light microscopy, wide-field microscopy; the use of magnification to visualize small things
Optic nerve	Neuroscience	Information highway connecting the retina to the thalamus (by way of the optic chiasm)
Optimization	Data science	The tuning, tweaking, fiddling, or systematic altering of parameters to maximize some desired outcome
os	Python	Package for operating system tools
Oscillation	Neuroscience, math	Movement to and fro; waves in the brain; likely not really that significant

(Continued)

	Type	Definition
Overfitting	Statistics	The use of too many variables, often resulting in a model's inability to generalize to datasets other than those used to build the model
p-hacking	Statistics	Bag of techniques (e.g., using multiple variables and flexible stopping rules) until $p < 0.05$ is reached; cheating
p-value	Statistics	Probability of obtaining data this extreme or more extreme assuming chance. $p < 0.05$ is often arbitrarily used as the threshold for whether or not a result is "unlikely enough" and a study should be published
Package	Python	Function or library of software
Parvalbumin	Neuroscience	PV, thought to label a particular class of fast-spiking inhibitory neurons
Passband	Engineering	Region of frequencies that a filter lets through, oftentimes suppressing other frequencies
Passing	CS	The act of sending a reference to an object to avoid sending the entire object
Passive	Engineering	Not active
Patch clamp	Neuroscience, engineering	Method for recording the activity of single neurons
PCA	Math, stats	See principal component analysis
Pdb	Python	Package useful for debugging (for example, in your code, place the text pdb. set_trace() on its own line and run the code, this will stop the code at this line and allow you to explore variables at this point).
Pearson-r correlation	Machine learning, statistics, neuroscience	A measure of the linear correlation between two variables on interval scale or higher
Periodic	Math	Occurring regularly or rhythmically
Perl	CS	Nobody uses Perl anymore
Phase	mathematics	Angle, or the starting point of wave
Photoreceptors	Neuroscience	Cells that transduce photons into electrical signals
Pickle	Python	Function for importing/exporting data

(Continued)

	Type	Definition
Pinna	Neuroscience	Outer ear
Pipeline	Engineering	The computation machinery in place to handle a stream of data
Pivot	Industry	Change course; alter direction; critical in the development cycle of almost any startup
Poisson	Stats	A frequency distribution that gives the probability of temporally independent events; often incorrectly assumed to be the distribution of spike events. Initially used to model the number of people killed by horse kicks in the Prussian army from year to year
Population code	Neuroscience	Theory that groups of neurons encode and transmit signals through an aggregate spiking method
Potassium	Neuroscience	K (for Kalium), important ion for membrane potential balance and spiking
Power	Engineering	Amplitude or magnitude squared
Precognition	Neuroscience	Method of seeing into the future; a lie
Presbycusis	Neuroscience, medicine	Age-related hearing loss
Presbyopia	Neuroscience, medicine	Age-related vision loss
Primary auditory cortex	Neuroscience	Principal input structure in the cortex for information about sounds in the world
Principal component analysis	Math	PCA, method for reducing the number of dimensions of a data set to something smaller and more manageable
Projection neuron	Neuroscience	Neuron that sends an axon to another region, typically far away
PSTH	Neuroscience	Peri-stimulus time histogram, means for visualizing neural activity around a repeated stimulus
Psychophysics	Neuroscience, psychology	A field that tries to find functional relationships between physical stimuli and the mental perception of these stimuli, pioneered by Gustav Theodor Fechner in the 1860s
Pumps	Neuroscience	Membrane structures used to move ions in or out of a cell
Purkinje cell	Neuroscience	Large cell in the cerebellum, with very big dendritic arbor

(Continued)

	Type	Definition
Pyramidal neuron	Neuroscience	Large cortical neurons shaped like pyramids, typically excitatory
R	Computing	Popular language for statistical computing
RAM	Computing	Random access memory
Random Forest	Machine learning	Method that uses multiple decision trees for regression or classification
Rate code	Neuroscience	Theory that neural messages are transmitted as spikes per time interval—in opposition to the notion that there are specific motifs of activity that matter
Recommendation Engine	Data science	Method for personalized product recommendations based on similarities to other users, popularized by Amazon and Netflix
Regression	Statistics	General term for trying to predict the value of one variable by knowing the value of one or more other variables, based on correlation, pioneered by Galton
Regularization	ML	Method for preventing unreasonably variable coefficients to prevent overfitting
Reinforcement Learning	CS, data science, neuroscience, ML	Field of study dedicated to understanding how decisions are made based on memory of reward history
Resting potential	Neuroscience	Voltage at which membrane voltage is stable (nonfiring)
Resting state	Neuroscience	A neuron that is stably not spiking and not receiving considerable excitatory or inhibitory input so as to significantly alter its membrane potential
Retina	Neuroscience	The tissue in the back of the eye that transforms photons to electrical signals
Retinal ganglion cells	Neuroscience	Neurons in the retina of the eye that project to the brain, and the axons of which form the optic nerve
Reward prediction error	Neuroscience	RPE, when one expects a reward but doesn't get it
Rig	Neuroscience, engineering	Entire apparatus used to collect data, including amplifiers, headstages, etc.
Rippling	Engineering	Residual variation, e.g. in the side lobes of a filter
ROC curve	Statistics	Receiver-operating characteristic curve, method from signal detection theory
Rod	Neuroscience	Photoreceptor used for monochromatic vision at low light levels (night vision), dominates in the retinal periphery

(*Continued*)

	Type	Definition
Sata	Data Science, Neuroscience	The quantified result of a systematic logging, capturing information about simulations. Different from data, but sometimes a useful approximation
Sataset	Data Science, Neuroscience	A collection of sata
Scalar	Math	Single number
Scatter	Python (matplotlib.pyplot), MATLAB	Method for visualizing data as points, usually in a two-dimensional plot, e.g., two variables against each other
Scientific method	Life	Method for the systematic generation of knowledge (Wissenschaft), usually by systematic data collection to test hypotheses
Scipy	Python	Python package for scientific computing
Seaborns	Python	Python package for good-looking and formatted visualization
Side lobes	Engineering	Frequency-domain bands in a filter that are not part of the main lobe or pass band
Simulation	Neuroscience, computing	A (usually simplified) computer model of a natural system, used to make predictions that need to be tested empirically
Sinc Function	Engineering	Sine Cardinal; a sine-like wave whose magnitude is inversely proportional to x
Slepian Function	Engineering	Prolate Spheroidal Wave Function.
SNR	Engineering	Signal to noise ratio: the magnitude of the signal divided by the magnitude of the noise
Society for Neuroscience	Neuroscience	A large group of neuroscientists who convene annually at a big conference known as SFN or Neuroscience. They also publish the *Journal of Neuroscience*
Sodium	Neuroscience	Ion critical for voltage regulation in neurons, particularly relevant for spike initiation
Space clamp	Neuroscience	Method for collapsing the virtual length of neuronal process to zero; funk rocker
Spark	CS	Programming language built for distributed processing
Sparkulum	NDS	Smallest unit of useful intelligence learned that propagates more learning

(Continued)

	Type	Definition
Spectrogram	Engineering	Visualization of power at different frequencies of a signal within a certain time window, over time
Spike threshold	Neuroscience	The membrane voltage at which a spike is initiated in a neuron
Spike triggered average	Neuroscience	STA, the average stimulus that preceded spiking activity
Spikes	Neuroscience	Binary events that presumably form the basic units of the neural code, as this is the signal that is exchanged by neurons over long distances
Spine (dendritic)	Neuroscience	Small projection from a dendrite, typically for forming synapses. These are quite dynamic, can grow on a timescale of hours, wither from stress
Spiny neuron	Neuroscience	Neuron that contains many spines
SQL	Industry, Data Science	Programming language for relational databases
Square wave	Engineering	[0, 0, 0, 0, 0, 0, 1, 1, 1, 1, 1, 1, 1, 1, 1, 0, 0, 0, 0, 0, 0]
Squid giant axon	Neuroscience	Large neuron axon that was the model system for pioneering neuroscientists like Hodgkin and Huxley
Stack exchange	CS	Website/community that is a great help for asking and answering questions related to programming
Stack overflow	CS	Website/community that is a great help for asking and answering questions related to programming
Stationary	Engineering	Unchanging in time.
Sublime	CS	Text editor, useful for programming in many different languages
Subplot	MATLAB, Python	A smaller figure within a larger figure
Sulcus	Neuroscience	Cortical valley
Supervised learning	Machine learning	Method for creating predictions based on known groupings/labels of data
SVD	Machine learning	Singular Value Decomposition, method of matrix manipulation
Synapse	Neuroscience	The point of connection between two neurons, the chemical interface for signal exchange between neurons

(Continued)

	Type	**Definition**
sys	Python	Library in Python useful for system level functions
t-test	Stats	Popular statistical method for testing whether two groups of data are likely drawn from the same distribution, particularly when sample size is small and population variance is unknown, pioneered by Gosset
TDCS	Neuroscience	See transcranial direction current stimulation
Temporal	Neuroscience	Related to time; also a lobe on the flank of the brain that houses the auditory system and memory systems
Terminal	Neuroscience, medicine	Deadly; the end
Tetanus	Neuroscience	Fast and repetitive; typical stimulus used to study LTD and LTP in hippocampal slices
Tetrode	Neuroscience, engineering	Collection of four electrodes useful for isolating single units (neurons) as the known architecture allows to triangulate waveforms
Tetrodotoxin	Neuroscience	TTX, inhibits firing of action potentials, very strong poison
Thalamus	Neuroscience	The main gateway of sensory information from the brainstem to the cortex. All sensory signals (except for smell) have to go through the thalamus to reach cortex
Theta wave	Neuroscience	Waves commonly measured in EEG near 6 Hz. Typical for stage 2 sleep
Thunder	Python	Python package developed by Jeremy Freeman for large-scale neural data analysis
Time constant	Neuroscience, math	A variable used often to describe the rate of activity decay for a particular system (e.g., voltage, radioactivity)
Timing	Neuroscience	Is everything
Timing code	Neuroscience	The theory that the precise time of neural activity represents information
Tinnitus	Neuroscience	The phantom phenomenon of "ringing" in the ears in the absence of an auditory stimulus
Tonotopy	Neuroscience	Principle of frequency-based organization in the auditory nervous system in analogy to retinotopy of the visual system
Tract cell	Neuroscience	Jargon to describe neurons that project their axons far away, e.g., "we labeled tract cells in the left part of the brain and saw signals in cell bodies on the right side"

(Continued)

	Type	Definition
Transcranial direction current stimulation	Neuroscience	The act of passing current through the brain through external electrodes; highly experimental
Transient	Engineering	Brief, not sustained
Tungsten	Neuroscience, engineering	Material used for electrodes to record extracellular spiking activity, pioneered by David Hubel
Tuning curve	Neuroscience	Similar to receptive field, a visualization of the response properties of a neuron for different stimuli, e.g., lines oriented at different angles
Tuple	Python	An immutable object (once you create it, you cannot change it, regardless of irresistible force)
Twinx	Python	Matplotlib function for creating a second axis, useful for plotting overlaying y-axes
Two alternative forced choice	Neuroscience	2AFC; type of experimental paradigm with two answer options in which the participant must select one of the two options
Two photon microscopy	Neuroscience, Engineering, Physics	Method of using a pulsed laser for imaging neural activity of deep tissue
Type 1 error	Statistics	False positive; incorrecting identifying something as true when it is false (also alpha-error)
Type 2 error	Statistics	False negative; incorrectly identifying something as false when it is true (also beta-error or miss)
Unit testing	CS, Industry	The practice of building code within a framework that tests for bugs
Univariate distribution	Stats	The probability distribution of one variable (in contrast to multivariate distributions that arise from multiple variables)
Unsupervised learning	Machine learning	Class of methods for predicting groups based on no ground truth of which groups the data belong to
Urllib2	Python	Python package for web-scraping
Urllib3	Python	Python package for web-scraping

(Continued)

	Type	**Definition**
V1	Neuroscience	Primary visual cortex or striate cortex—the largest cortical area associated with visual processing
V2	Neuroscience	Secondary visual cortex, thought to process image textures, figure/ground separation, and illusory contours
V4	Neuroscience	Another visual area, implicated in mid-level vision, processing of color and shape—part of the ventral stream
Van Rossum, Guido	Python	Creator of the computing language Python
Vector	MATLAB, math	A one-dimensional array
Vim	Computing	Text editor for writing computer code
Visual cortex	Neuroscience	Region of neocortex dedicated to the processing and representation of visual information
Visual field	Neuroscience	Region in space in which an observer can see objects if they are presented
Vitals	Neuroscience	Essential systems to monitor during an animal experiment
Voltage	Neuroscience, engineering, physics	The relative difference in charge between two regions, in neuroscience a measure of the state of a neural membrane
Voltage clamp	Neuroscience	Method for fixing/holding the voltage of a neuron's membrane a constant level to measure current
Voltage-gated channel	Neuroscience	Channel in a cell's membrane that usually opens in response to reaching a certain voltage, e.g., voltage-gated sodium channels are critical for the generation of action potentials
Wavelet	Engineering	Symmetric filter used for identifying patterns in signals (see Mike X Cohen's book for helpful discussion), signal goes from 0 to 0. What matters is what happens in between
Webscraper	CS, data science	A tool for gathering data from a website, typically in the absence of an API
where	Python (numpy)	Function for finding the index of occurrence of something

(*Continued*)

	Type	Definition
White matter	Neuroscience	Typically bundles or large areas of brain composed of axons, seen as the cables that connect brain areas
Whole-cell recording	Neuroscience	Method for recording from a cell intracellularly while keeping it alive
Wvtool	MATLAB	Function for viewing frequency and time domain characteristics of windows
Yellow fluorescent protein	Neuroscience	Protein that glows yellow, can be used as a label for visualizing neurons
YFP	Neuroscience	See yellow fluorescent protein

Bibliography

Abbott, B. P., Abbott, R., Abbott, T. D., Abernathy, M. R., Acernese, F., Ackley, K., ... Adya, V. B. (2016). Observation of gravitational waves from a binary black hole merger. *Physical Review Letters, 116*(6), 061102.

Abegglen, L. M., Caulin, A. F., Chan, A., Lee, K., Robinson, R., Campbell, M. S., ... Jensen, S. T. (2015). Potential mechanisms for cancer resistance in elephants and comparative cellular response to DNA damage in humans. *JAMA, 314*(17), 1850–1860.

Abrams, R. M., Griffiths, S. K., Huang, X., Sain, J., Langford, G., & Gerhardt, K. J. (1998). Fetal music perception: The role of sound transmission. *Music Perception: An Interdisciplinary Journal, 15*(3), 307–317.

Adrian, D., Bhargavan, K., Durumeric, Z., Gaudry, P., Green, M., Halderman, J. A., ... VanderSloot, B. (2015, October). Imperfect forward secrecy: How Diffie-Hellman fails in practice. In *Proceedings of the 22nd ACM SIGSAC conference on computer and communications security.* (pp. 5–17). Denver, Colorado, USA: ACM.

Adrian, E. D. (1926). The impulses produced by sensory nerve endings. *The Journal of Physiology, 61*(1), 49–72.

Akaike, H. (1974). A new look at the statistical model identification. *IEEE Transactions on Automatic Control, 19*(6), 716–723.

Akaike, H. (2011). Akaike's information criterion. In *International encyclopedia of statistical science.* (p. 25). Berlin Heidelberg: Springer.

Aramchek, U. [The Patanoiac]. (2014, Feb 17). An inquiry into neurometeorology and the discovery that hurricanes live in constant agony [Tweet]. Retrieved from https://twitter.com/thepatanoiac/status/435439436747137025.

Aristotle, , Jowett, B., & Davis, H. W. C. (1920). *Aristotle's politics.* Oxford: Clarendon Press.

Arrhenius, S. (1889). *Über die Dissociationswärme und den Einfluss der Temperatur auf den Dissociationsgrad der Elektrolyte.* Leipzig: Wilhelm Engelmann.

Arrhenius, S. (1889). Über die Reaktionsgeschwindigkeit bei der Inversion von Rohrzucker durch Säuren. *Zeitschrift für physikalische Chemie, 4,* 226–248.

Averbeck, B. B., Latham, P. E., & Pouget, A. (2006). Neural correlations, population coding and computation. *Nature Reviews Neuroscience, 7*(5), 358–366.

Baron Fourier, J. B. J. (1878). *The analytical theory of heat.* Cambridge: The University Press.

Borges, J. L. (1962). The garden of forking paths. Collected Fictions, New York, NY: Grove Press. pp. 119–128.

Boser, B. E., Guyon, I. M., & Vapnik, V. N. (1992, July). A training algorithm for optimal margin classifiers. In *Proceedings of the fifth annual workshop on computational learning theory.* (pp. 5–17). Pittsburgh: ACM.

Boyden, E. S., Zhang, F., Bamberg, E., Nagel, G., & Deisseroth, K. (2005). Millisecond-timescale, genetically targeted optical control of neural activity. *Nature Neuroscience, 8*(9), 1263–1268.

Bracewell, R. (1965). *The Fourier transform and it is applications.* New York: McGraw-Hill Book Co.

Breiman, L. (2001). Random forests. *Machine Learning, 45*(1), 5–32.

Brette, R., & Gerstner, W. (2005). Adaptive exponential integrate-and-fire model as an effective description of neuronal activity. *Journal of Neurophysiology, 94*(5), 3637–3642.

Brzychczy, S., & Poznanski, R. R. (2013). *Mathematical neuroscience.* Waltham, MA: Academic Press.

Buhr, E. D., Yoo, S. H., & Takahashi, J. S. (2010). Temperature as a universal resetting cue for mammalian circadian oscillators. *Science, 330*(6002), 379–385.

Butterworth, S. (1930). On the theory of filter amplifiers. *Wireless Engineer, 7*(6), 536–541.

Buzsáki, G., & Draguhn, A. (2004). Neuronal oscillations in cortical networks. *Science, 304*(5679), 1926–1929.

Casey, S. D. (2012). Windowing systems for time-frequency analysis. *Sample Theory Signal Image Process, 11*(2-3), 221–251.

Cayco-Gajic, N. A., Zylberberg, J., & Shea-Brown, E. (2015). Triplet correlations among similarly tuned cells impact population coding. *Frontiers in Computational Neuroscience, 9,* 57.

Champollion, J.F. (1828). *Précis du système hiéroglyphique des anciens Égyptiens*. L'Imprimerie Royale.

Churchland, M. M., Byron, M. Y., Ryu, S. I., Santhanam, G., & Shenoy, K. V. (2006). Neural variability in premotor cortex provides a signature of motor preparation. *The Journal of Neuroscience*, 26(14), 3697–3712.

Cohen, M. R., & Kohn, A. (2011). Measuring and interpreting neuronal correlations. *Nature Neuroscience*, 14(7), 811–819.

Cohen, M. X. (2014). *Analyzing neural time series data: theory and practice*. Cambridge, MA: MIT Press.

Comon, P. (1994). Independent component analysis, a new concept? *Signal Processing*, 36(3), 287–314.

Cooley, J. W., & Tukey, J. W. (1965). An algorithm for the machine calculation of complex Fourier series. *Mathematics of Computation*, 19(90), 297–301.

Costa, P.T., & McCrae, R.R. (1985). *The NEO personality inventory*.

Costa, P. T., & McCrae, R. R. (1992). Four ways five factors are basic. *Personality and Individual Differences*, 13(6), 653–665.

Craft, E., Schütze, H., Niebur, E., & Von Der Heydt, R. (2007). A neural model of figure–ground organization. *Journal of Neurophysiology*, 97(6), 4310–4326.

Crumiller, M., Knight, B., Yu, Y., & Kaplan, E. (2011). Estimating the amount of information conveyed by a population of neurons. *Frontiers in Neuroscience*, 5, 90.

Dennett, D. C. (1993). *Consciousness explained*. London: Penguin.

Dorn, J. D., & Ringach, D. L. (2003). Estimating membrane voltage correlations from extracellular spike trains. *Journal of Neurophysiology*, 89(4), 2271–2278.

Doyle, D. A., Cabral, J. M., Pfuetzner, R. A., Kuo, A., Gulbis, J. M., Cohen, S. L., … MacKinnon, R. (1998). The structure of the potassium channel: Molecular basis of K+ conduction and selectivity. *Science*, 280(5360), 69–77.

Dyson, F. (2004). A meeting with Enrico Fermi. *Nature*, 427(6972), 297.

Ecker, A. S., Berens, P., Keliris, G. A., Bethge, M., Logothetis, N. K., & Tolias, A. S. (2010). Decorrelated neuronal firing in cortical microcircuits. *Science*, 327(5965), 584–587.

Efron, B., & Tibshirani, R. J. (1994). *An introduction to the bootstrap*. Boca Raton, FL: CRC Press.

Euler, L. (1748). *Introductio in analysin infinitorum* (Vol. 2). Lausannae: MM Bousquet.

Fechner, G. T. (1860). *Elements of psychophysics*. Leipzig: Breitkopf & Härtel.

Fourier, J. (1822). *Théorie analytique de la chaleur*. Paris: Firmin Didot Père et Fils.

Frégnac, Y., & Laurent, G. (2014). Neuroscience: where is the brain in the human brain project. *Nature*, 513(7516), 27–29.

Fries, P. (2005). A mechanism for cognitive dynamics: neuronal communication through neuronal coherence. *Trends in Cognitive Sciences*, 9(10), 474–480.

Fries, P. (2009). Neuronal gamma-band synchronization as a fundamental process in cortical computation. *Annual Review of Neuroscience*, 32, 209–224.

Fyfe, S., Williams, C., Mason, O. J., & Pickup, G. J. (2008). Apophenia, theory of mind and schizotypy: Perceiving meaning and intentionality in randomness. *Cortex*, 44(10), 1316–1325.

Galton, F. (1886). Regression towards mediocrity in hereditary stature. *The Journal of the Anthropological Institute of Great Britain and Ireland*, 15, 246–263.

Gegenfurtner, K. R., & Kiper, D. C. (2003). Color vision. *Annual Review of Neuroscience*, 26(1), 181–206.

Gladwell, M. (2002). Blowing up. *The New Yorker*, 162.

Gómez, D. M., Berent, I., Benavides-Varela, S., Bion, R. A., Cattarossi, L., Nespor, M., & Mehler, J. (2014). Language universals at birth. *Proceedings of the National Academy of Sciences*, 111(16), 5837–5841.

Good, I. J. (1979). Studies in the history of probability and statistics. XXXVII AM Turing's statistical work in World War II. *Biometrika*, 393–396.

Green, P., & Yavin, E. (1998). Mechanisms of docosahexaenoic acid accretion in the fetal brain. *Journal of Neuroscience Research*, 52(2), 129–136.

Gürlebeck, N. (2015). No-hair theorem for black holes in astrophysical environments. *Physical Review Letters*, 114(15), 151102.

Gutnisky, D. A., & Dragoi, V. (2008). Adaptive coding of visual information in neural populations. *Nature*, 452(7184), 220–224.

Heffner, R. S., & Heffner, H. E. (1985). Hearing range of the domestic cat. *Hearing Research*, 19(1), 85–88.

Hilbert, M., & López, P. (2011). The world's technological capacity to store, communicate, and compute information. *Science*, 332(6025), 60–65.

Hodgkin, A. L., & Huxley, A. F. (1952). A quantitative description of membrane current and its application to conduction and excitation in nerve. *The Journal of Physiology*, 117(4), 500.

Horn, J. L. (1965). A rationale and test for the number of factors in factor analysis. *Psychometrika*, *30*(2), 179–185.

Hubel, D. H., & Wiesel, T. N. (1968). Receptive fields and functional architecture of monkey striate cortex. *The Journal of Physiology*, *195*(1), 215–243.

Hubel, D. H., & Wiesel, T. M. (2004). *Brain and visual perception: The story of a 25-year collaboration*. New York, NY: Oxford University Press.

Huk, A. C., & Shadlen, M. N. (2005). Neural activity in macaque parietal cortex reflects temporal integration of visual motion signals during perceptual decision making. *The Journal of Neuroscience*, *25*(45), 10420–10436.

Jakab, P. L. (2014). *Visions of a flying machine: The Wright brothers and the process of invention*. Washington and London: Smithsonian Institution.

James, W. (1890). *The principles of psychology*. New York, NY: Read Books Ltd.

Jazayeri, M., Wallisch, P., & Movshon, J. A. (2012). Dynamics of macaque MT cell responses to grating triplets. *The Journal of Neuroscience*, *32*(24), 8242–8253.

Jonas, E., & Kording, K. (2016). Could a neuroscientist understand a microprocessor? *bioRxiv*, 055624.

Kajikawa, Y., & Schroeder, C. E. (2011). How local is the local field potential? *Neuron*, *72*(5), 847–858.

Kaufman, M. T., Churchland, M. M., Ryu, S. I., & Shenoy, K. V. (2014). Cortical activity in the null space: Permitting preparation without movement. *Nature Neuroscience*, *17*(3), 440–448.

Koch, C. (2004). *Biophysics of computation: Information processing in single neurons*. New York: Oxford University Press.

Kohn, A., & Smith, M. A. (2005). Stimulus dependence of neuronal correlation in primary visual cortex of the macaque. *The Journal of Neuroscience*, *25*(14), 3661–3673.

Kumar, S. (2012). Fundamental limits to Moore's law. *Fundamental Limits to Moore's Law*, 9, . *Stanford University*.

Kurzweil, R. (2005). *The singularity is near: When humans transcend biology*. New York: Penguin.

Kurzweil, R., Richter, R., & Schneider, M. L. (1990). In *The age of intelligent machines* (Vol. 579). Cambridge: MIT Press.

Latimer, K. W., Yates, J. L., Meister, M. L., Huk, A. C., & Pillow, J. W. (2015). Single-trial spike trains in parietal cortex reveal discrete steps during decision-making. *Science*, *349*(6244), 184–187.

Ledesma, R. D., & Valero-Mora, P. (2007). Determining the number of factors to retain in EFA: An easy-to-use computer program for carrying out parallel analysis. *Practical Assessment, Research & Evaluation*, *12*(2), 1–11.

Leibniz, G.W. (1714). *La Monadologie*, edition établie par E. Boutroux, Paris LGF.

Levy, R. B., & Reyes, A. D. (2011). Coexistence of lateral and co-tuned inhibitory configurations in cortical networks. *PLoS Computational Biology*, *7*(10), e1002161.

Lewicki, M. S. (1998). A review of methods for spike sorting: The detection and classification of neural action potentials. *Network: Computation in Neural Systems*, *9*(4), R53–R78.

Lintott, C. J., Schawinski, K., Slosar, A., Land, K., Bamford, S., Thomas, D., … Murray, P. (2008). Galaxy Zoo: Morphologies derived from visual inspection of galaxies from the Sloan Digital Sky Survey. *Monthly Notices of the Royal Astronomical Society*, *389*(3), 1179–1189.

Liu, J., Li, J., Feng, L., Li, L., Tian, J., & Lee, K. (2014). Seeing Jesus in toast: Neural and behavioral correlates of face pareidolia. *Cortex*, *53*, 60–77.

Lockery, S. R., & Goodman, M. B. (2009). The quest for action potentials in *C. elegans* neurons hits a plateau. *Nature Neuroscience*, *12*(4), 377–378.

Logothetis, N. K., Pauls, J., Augath, M., Trinath, T., & Oeltermann, A. (2001). Neurophysiological investigation of the basis of the fMRI signal. *Nature*, *412*(6843), 150–157.

Lomborg, B. (2003). *The skeptical environmentalist: Measuring the real state of the world* (Vol. 1). Cambridge: Cambridge University Press.

Mainen, Z. F., & Sejnowski, T. J. (1995). Reliability of spike timing in neocortical neurons. *Science*, *268*(5216), 1503.

Malthus, T. R. (1826). *An essay on the principle of population*. London: W. Pickering.

Mante, V., Sussillo, D., Shenoy, K. V., & Newsome, W. T. (2013). Context-dependent computation by recurrent dynamics in prefrontal cortex. *Nature*, *503*(7474), 78–84.

Markram, H. (2006). The blue brain project. *Nature Reviews Neuroscience*, *7*(2), 153–160.

Markram, H. (2012). The human brain project. *Scientific American*, *306*(6), 50–55.

Marmarelis, P. Z., & Marmarelis, V. Z. (1978). *Analysis of physiological systems: The white-noise approach*. New York, NY: Plenum Press.

Mazurek, M. E., Roitman, J. D., Ditterich, J., & Shadlen, M. N. (2003). A role for neural integrators in perceptual decision making. *Cerebral Cortex, 13*(11), 1257–1269.

McCrae, R. R., & Costa, P. T. (1987). Validation of the five-factor model of personality across instruments and observers. *Journal of Personality and Social Psychology, 52*(1), 81.

McCrae, R. R., & Costa, P. T. (1997). Personality trait structure as a human universal. *American Psychologist, 52*(5), 509.

McCrae, R. R., & Costa, P. T. (2003). *Personality in adulthood: A five-factor theory perspective.* New York: Guilford Press.

Meadows, D. H., Meadows, D. L, Randers, J., & Behrens, W. W. (1972). *The limits to growth.* (p. 102). New York: Universe Books.

Mechler, F., & Ringach, D. L. (2002). On the classification of simple and complex cells. *Vision Research, 42*(8), 1017–1033.

Mishkin, M., Ungerleider, L. G., & Macko, K. A. (1983). Object vision and spatial vision: two cortical pathways. *Trends in Neurosciences, 6,* 414–417.

Mitchell, J. F., Sundberg, K. A., & Reynolds, J. H. (2007). Differential attention-dependent response modulation across cell classes in macaque visual area V4. *Neuron, 55*(1), 131–141.

Mitra, P. (2007). *Observed brain dynamics.* New York: Oxford University Press.

Moore, G. E. (2006). Cramming more components onto integrated circuits, Reprinted from Electronics, volume 38, number 8, April 19, 1965, pp. 114 ff. *IEEE Solid-State Circuits Newsletter, 3*(20), 33–35.

Movshon, J. A., Adelson, E. H., Gizzi, M. S., & Newsome, W. T. (1992). *The analysis of moving visual patterns.* Cambridge, MA: MIT Press.

Nagel, T. (1974). What is it like to be a bat? *The Philosophical Review, 83*(4), 435–450.

Nagy, J. D., Victor, E. M., & Cropper, J. H. (2007). Why don't all whales have cancer? A novel hypothesis resolving Peto's paradox. *Integrative and Comparative Biology, 47*(2), 317–328.

Newton, I. (1687). *Philosophiae naturalis principia mathematica.* Londini: Jussu Societatis Regiae ac typis Iosephi Streater: Prostat apud plures bibliopolas.

Neyman, J. (1937). Outline of a theory of statistical estimation based on the classical theory of probability.. *Philosophical Transactions of the Royal Society of London. Series A, Mathematical and Physical Sciences, 236*(767), 333–380.

Nienborg, H., & Cumming, B. G. (2009). Decision-related activity in sensory neurons reflects more than a neuron's causal effect. *Nature, 459*(7243), 89–92.

Nirenberg, S. H., & Victor, J. D. (2007). Analyzing the activity of large populations of neurons: How tractable is the problem? *Current Opinion in Neurobiology, 17*(4), 397–400.

Nylen, E. L. (2016). *The cortical transformation of sound envelope in the mouse.* Dissertation. New York University.

Ohiorhenuan, I. E., & Victor, J. D. (2011). Information-geometric measure of 3-neuron firing patterns characterizes scale-dependence in cortical networks. *Journal of Computational Neuroscience, 30*(1), 125–141.

Ohiorhenuan, I. E., Mechler, F., Purpura, K. P., Schmid, A. M., Hu, Q., & Victor, J. D. (2010). Sparse coding and high-order correlations in fine-scale cortical networks. *Nature, 466*(7306), 617–621.

Olshausen, B. A., & Field, D. J. (2004). What is the other 85% of V1 doing. *Problems in Systems Neuroscience, 4*(5), 182–211.

Olshausen, B. A., & Field, D. J. (2005). How close are we to understanding V1? *Neural Computation, 17*(8), 1665–1699.

Open Science Collaboration.(2015). Estimating the reproducibility of psychological science. *Science, 349*(6251), aac4716.

Peters, T. (2004). PEP 20 -- The Zen of Python. Retrieved from https://www.python.org/dev/peps/pep-0020/.

Pillow, J. W., Shlens, J., Chichilnisky, E. J., & Simoncelli, E. P. (2013). A model-based spike sorting algorithm for removing correlation artifacts in multi-neuron recordings. *PLoS One, 8*(5), e62123.

Pinker, S. (1999). How the mind works. *Annals of the New York Academy of Sciences, 882*(1), 119–127.

Planck, M. (1900). Über irreversible Strahlungsvorgänge. *Annalen der Physik, 306*(1), 69–122.

Querleu, D., Renard, X., Versyp, F., Paris-Delrue, L., & Crèpin, G. (1988). Fetal hearing. *European Journal of Obstetrics & Gynecology and Reproductive Biology, 28*(3), 191–212.

Rall, W. (1969). Time constants and electrotonic length of membrane cylinders and neurons. *Biophysical Journal, 9*(12), 1483.

Reich, D. S., Mechler, F., & Victor, J. D. (2001). Independent and redundant information in nearby cortical neurons. *Science, 294*(5551), 2566–2568.

Ricardo, D. (1891). *Principles of political economy and taxation.* London: G. Bell and Sons.

Rieke, F. (1999). *Spikes: Exploring the neural code*. Cambridge, MA: MIT Press.

Rimfeld, K., Kovas, Y., Dale, P. S., & Plomin, R. (2016). True grit and genetics: Predicting academic achievement from personality. *Journal of Personality and Social Psychology*. Retrieved from http://dx.doi.org/10.1037/pspp0000089.

Rojas-Líbano, D., & Kay, L. M. (2008). Olfactory system gamma oscillations: The physiological dissection of a cognitive neural system. *Cognitive Neurodynamics*, 2(3), 179–194.

Rousseeuw, P. J. (1987). Silhouettes: A graphical aid to the interpretation and validation of cluster analysis. *Journal of Computational and Applied Mathematics*, 20, 53–65.

Salinas, E., & Sejnowski, T. J. (2001). Correlated neuronal activity and the flow of neural information. *Nature Reviews Neuroscience*, 2(8), 539–550.

Salsburg, D. (2001). *The lady tasting tea: How statistics revolutionized science in the twentieth century*. New York: Macmillan.

Shadlen, M. N., & Movshon, J. A. (1999). Synchrony unbound: A critical evaluation of the temporal binding hypothesis. *Neuron*, 24(1), 67–77.

Shannon, C. (1948). A mathematical theory of communication. *Bell System Technical Journal*, 27, 379–423. 623-656.

Shermer, M. (2008). Patternicity: Finding meaningful patterns in meaningless noise. *Scientific American*, 299(5), 48.

Simon, J. L. (1998). *The ultimate resource*. Princeton: Princeton University Press. p. 2.

Simons, J., Nelson, L., & Simonsohn, U. (2011). Undisclosed flexibility in data collection and analysis allows presenting anything as significant. *Psychological Science*, 22, 1359–1366.

Smith, M. A., & Kohn, A. (2008). Spatial and temporal scales of neuronal correlation in primary visual cortex. *The Journal of Neuroscience*, 28(48), 12591–12603.

Snyder, A. C., Morais, M. J., Willis, C. M., & Smith, M. A. (2015). Global network influences on local functional connectivity. *Nature Neuroscience*, 18(5), 736–743.

Softky, W. R., & Koch, C. (1993). The highly irregular firing of cortical cells is inconsistent with temporal integration of random EPSPs. *The Journal of Neuroscience*, 13(1), 334–350.

Stevenson, I. H., & Kording, K. P. (2011). How advances in neural recording affect data analysis. *Nature Neuroscience*, 14(2), 139–142.

Stigler, S. M. (2016). *The seven pillars of statistical wisdom*. Cambridge, MA: Harvard University Press.

Taleb, N. N. (2007). *The black swan: The impact of the highly improbable*. New York: Random House Incorporated.

Taleb, N. N. (2012). *Antifragile: Things that gain from disorder* (Vol. 3). New York: Random House Incorporated.

Taouali, W., Benvenuti, G., Wallisch, P., Chavane, F., & Perrinet, L. U. (2016). Testing the odds of inherent vs. observed overdispersion in neural spike counts. *Journal of Neurophysiology*, 115(1), 434–444.

Theibault, J. (1997). The demography of the thirty years war re-revisited: Günther Franz and his Critics. *German History*, 15(1), 1–21.

Taylor, B. N., & Thompson, A. (2008). *The International System of Units (SI)*. Special Publication 330. Gaithersburg, MD: National Institute of Standards and Technology.

Tierney, B. (1997). In *The idea of natural rights: Studies on natural rights, natural law, and church law* (Vol. 5). Rapids, MI: Wm. B. Eerdmans Publishing. pp. 1150–1625.

Toomer, G. J. (1984). In *Ptolemy's almagest*, 1984. New York: Springer-Verlag. p. 1.

Tukey, J. W. (1977). *Exploratory data analysis*. Princeton, NJ: Princeton University Press.

Turing, A. M. (1938). On computable numbers, with an application to the entscheidungsproblem. *Journal of Mathematics*, 58, 345–363.

Tye, K. M., & Deisseroth, K. (2012). Optogenetic investigation of neural circuits underlying brain disease in animal models. *Nature Reviews Neuroscience*, 13(4), 251–266.

Uhlhaas, P., Pipa, G., Lima, B., Melloni, L., Neuenschwander, S., Nikolić, D., & Singer, W. (2009). Neural synchrony in cortical networks: History, concept and current status. *Frontiers in Integrative Neuroscience*, 3, 17.

Vadillo, M.A., Hardwicke, T.E., & Shanks, D.R. (2016). Selection bias, vote counting, and money-priming effects: A comment on Rohrer, Pashler, and Harris (2015) and Vohs (2015).

Van Kan, P. L. E., Scobey, R. P., & Gabor, A. J. (1985). Response covariance in cat visual cortex. *Experimental Brain Research*, 60(3), 559–563.

Von Uexküll, J. (1992). A stroll through the worlds of animals and men: A picture book of invisible worlds. *Semiotica*, 89(4), 319–391.

Waldrop, M. M. (2016). The chips are down for Moore's law. *Nature News*, 530(7589), 144.

Wallisch, P. (2015). An Expert's Lesson from the Dress. Slate. Retrieved from http://www.slate.com/articles/health_and_science/science/2016/03/the_science_of_the_black_and_blue_dress_one_year_later.html.

Wallisch, P. (2016a). Explaining Color Constancy. Creativity Post. Retrieved from http://www.creativitypost.com/science/explaining_color_constancy.

Wallisch, P. (2016b). The Meme That Spawned a Science Bonanza. Slate. Retrieved from http://www.slate.com/articles/health_and_science/science/2016/03/the_science_of_the_black_and_blue_dress_one_year_later.html.

Wallisch, P. (2011, January 7). Eponyms are stifling scientific progress [Web log post]. Retrieved from URL of blog post http://pensees.pascallisch.net/?p=686.

Wallisch, P. (2014, July 21). The relative scale of early visual areas [Web log post]. Retrieved from URL of blog post http://pensees.pascallisch.net/?p=1788.

Wallisch, P. (2014, July 30). What should we call simulated data? [Web log post]. Retrieved from URL of blog post http://pensees.pascallisch.net/?p=1801.

Wallisch, P. [Pascallisch]. (2013, October 2). People keep talking about "data science". Is there any other kind? [Tweet]. Retrieved from https://twitter.com/Pascallisch/status/385283767952105474.

Wallisch, P. [Pascallisch]. (2014, May 13). Every time you talk about "data science", you might as well say "I'm stupid". Is there a science without data? [Tweet]. Retrieved from https://twitter.com/Pascallisch/status/466343390947590144.

Wallisch, P., & Movshon, J. A. (2008). Structure and function come unglued in the visual cortex. *Neuron*, 60(2), 195–197.

Wallisch, P., Lusignan, M. E., Benayoun, M. D., Baker, T. I., Dickey, A. S., & Hatsopoulos, N. G. (2014). *MATLAB for neuroscientists: An introduction to scientific computing in MATLAB*. Academic Press.

Watson, M. (2013). Retrieved from http://opexanalytics.com/what-is-a-data-scientist/.

Webb, S. (2002). *If the universe is teeming with aliens... where is everybody?: Fifty solutions to the Fermi paradox and the problem of extraterrestrial life.* Springer Science & Business Media.

Weisberg, I., Tran, P., Christensen, B., Sibani, S., & Rozen, R. (1998). A second genetic polymorphism in methylenetetrahydrofolate reductase (MTHFR) associated with decreased enzyme activity. *Molecular Genetics and Metabolism*, 64(3), 169–172.

Wickelgren, I. (2006). Vision's grand theorist. *Science*, 314(5796), 78–79.

Xia, Y., Tong, H., Li, W. K., & Zhu, L. X. (2002). An adaptive estimation of dimension reduction space. *Journal of the Royal Statistical Society: Series B (Statistical Methodology)*, 64(3), 363–410.

Yeomans, K. A., & Golder, P. A. (1982). The Guttman-Kaiser criterion as a predictor of the number of common factors. *The Statistician*, 221–229.

Zohary, E., Shadlen, M. N., & Newsome, W. T. (1994). Correlated neuronal discharge rate and its implications for psychophysical performance. *Nature*, 370, 140–143.

Zukav, G. (2012). *The dancing Wu Li masters: An overview of the new physics.* New York: Random House Incorporated.

Index

Note: Page numbers followed by "*f*" and "*t*" refer to figures and tables, respectively.